LOST MINES and Historic TREASURES of British Columbia

by N. L. Barlee

hancock
house

ISBN 0-88839-992-8

Cover Photos by Murphy Shewchuk

Canadian Cataloguing in Publication Data

N. L. Barlee
Lost Mines and Historic Treasures of British Columbia
ISBN 0-88839-992-8
I. Title.

Printed in Hong Kong

Published simultaneously in Canada and the United States by

hancock house

HANCOCK HOUSE PUBLISHERS LTD.
19313 Zero Ave., Surrey, B.C. V3S 5J9

HANCOCK HOUSE PUBLISHERS
1431 Harrison Ave., Blaine, WA 98230

CONTENTS

FOREWORD

A rotting oak keg half full of gold and silver coins, a leather poke full of glinting nuggets, a long forgotten cache of old bank bills, burlap ore sacks loaded with high grade silver. These are but a few of the scores of treasures recovered in British Columbia over the past two decades, most of them, for usually logical reasons, unrecorded.

This book is concerned with some of the lost mines and treasures of British Columbia. There are, of course, innumerable stories of lost treasures and legendary mines; nearly every region in the province has one or more accounts pertaining to these fascinating subjects.

Unfortunately, most of these tales are just that. They are either highly embellished or simply erroneous. Our research, undertaken over a ten year period, indicates that only about seven out of every hundred stories in this category are worth investigating.

Some areas, however, have proven to be much more interesting than others. It is generally true that the historic mining regions like the Cariboo, Cassiar, Similkameen, West Kootenay, Fraser River and other old mining localities have yielded magnificent finds from time to time. The ghost towns have also been productive places and abandoned towns like Granite City, Sandon, Blakeburn, Cascade, Phoenix, Quesnelle Forks, Discovery and at least a dozen others have given up significant and often valuable treasures. The older towns and cities like Victoria, New Westminster, Nelson, Rossland, Yahk, Yale and scores of others have also indicated that the incidence of treasures in these historic places is much higher than in more recently established towns.

The publication of this work was not undertaken without a considerable amount of deliberation and perhaps some apprehension. To focus attention on a subject of this nature could be considered questionable but after postponing this publication for five years we finally decided to go ahead with the printing. Our reasons, we feel, are justified. After perusing all of the available facts we felt that there were no longer any valid reasons for supressing the material. Although we feel that any finds of historic treasures are part of our heritage unfortunately the laws of the province are so inadequate that they have failed consistently to protect or preserve this particular part of our past.

We also discovered, to our dismay, that a considerable number of our treasures, especially those consisting of specie, bullion and jewels, have been recovered by individuals who evidently experienced very little difficulty in transporting their finds across the border.

It is common knowledge that British Columbia has attracted a host of individuals loosely known as "pot hunters," a term used to describe a person who searches for treasure solely for monetary gain. This book, we hope, will help to eradicate this practice. We anticipate at least that more treasures will be retained in the province and hope that more adequate laws will be passed to protect this part of our heritage.

GENERAL LOCATIONS

Some of the historic treasures and lost mine sites are indicated in the general map below by the crosses.

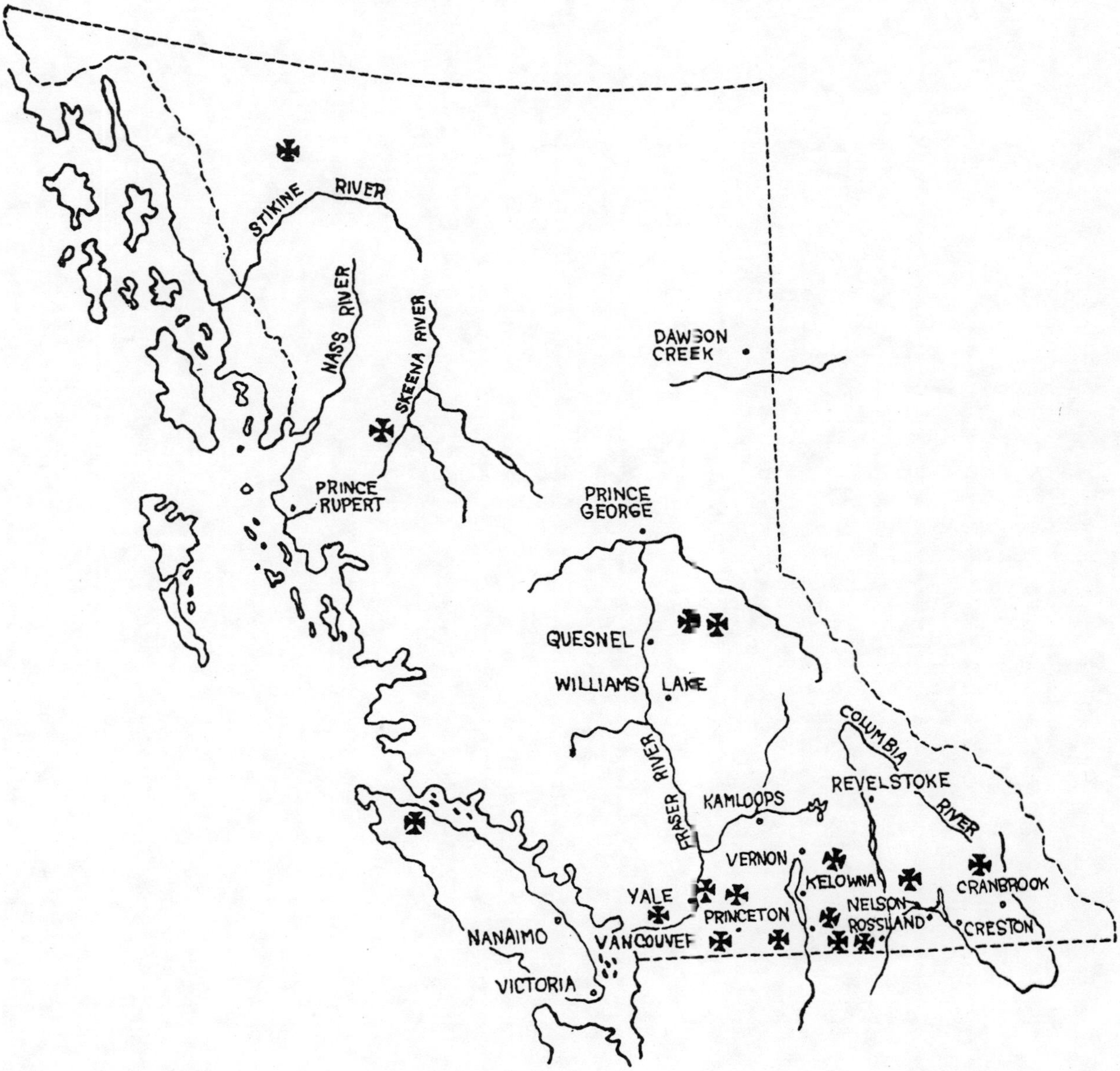

TREASURE

The Hotel Fairview in 1899. Known as the "Big Teepee," it burned to the ground in 1902. In the debris of this hotel coins, rings and other items of jewellery lost in the blaze at the time were recovered.

in HISTORIC FAIRVIEW

This is a story about a treasure find in the vanished town of Fairview in the arid south Okanagan country of southern British Columbia. This discovery, in 1969, consisted of jewellery, coins and other artifacts. This is a reprint from the Winter, 1969 issue of Canada West Magazine. The writer and an associate, C. W. Murray, made the discovery.

The first "Golden Gate" Hotel at Fairview. Built in 1892 by F. R. Kline, it was the first of five hotels in the mining town in the boom years.

In the west there were mining towns that appeared briefly on the scene but because of numerous factors never achieved their anticipated greatness and within a few decades, they slowly and almost imperceptibly slipped into obscurity and finally vanished. Such a town was Fairview in the arid south Okanagan desert country, a mining camp once balanced on the edge of greatness but now almost unknown outside the annals of Canadian mining.

There remains little to mark the passing of this once vaunted camp: the original town has virtually disappeared, the great mines have long been closed and nature has moved in to reclaim its own on the sagebrush covered flats. Today, a stranger would simply pass by for a casual glance reveals little to excite the imagination, and Fairview seems no different than a host of other mining camps which have undergone a similar climb and inevitable fall. There are, however, ghost towns that guard a secret well and then, at a most unexpected moment, reveal their bonanza. Fairview was this type of ghost town.

In the early 1880s, "One Armed" Reed and his Irish partner, roving through the south Okanagan hills, unsuccessfully panned for gold in an unnamed little mountain stream which flowed east into the Okanagan River in the valley far below. They were strictly placer miners. Thus when the gravels failed to yield colours they moved on, Reed pausing to leave only his name to the little creek in the gulch, which unknown to them, contained a vast body of gold bearing ore. In time the lode prospectors followed and in 1887, Sheenan and Gwatkins staked a discovery claim on what was later to become the renowned Stemwinder Mine. It was a rich find with free gold visible in the quartz. The result was a rush to the area. By 1898, the Morning Star, Tin Horn, Wild Horse, Joe Dandy and Brown Bear were all producing and Fairview was fast becoming established as a mining town with a promise of long life. F.R. Kline was convinced of its future and built the two story "Golden Gate" Hotel in 1892. Others were soon to follow; the Miner's Rest, Jack Moffat's log saloon far above the townsite, the "Fish House" which was also called the Second Golden Gate and finally in 1898, the grandest of them all was built. It was the Hotel Fairview, later to be known as the "Big Teepee," an elegant three story structure reputed to be the finest hotel in the interior. Fairview had arrived. It became a regular stop for the stagecoaches on their run from Oro (Oroville) to Penticton. Day in and day out the freighting teams of the Bassett Brothers strained up the steep grades with machinery for the busy mines and the town became a magnet for promoters from all corners of the earth. Down its dusty streets strode men like "Father Pat," the wandering Anglican minister known as "The Miner's Priest," Bill Miner, train robber without equal, strange Matthew Roderick, civil engineer by trade and hold-up man by preference and a myriad of others. As Fairview's fame spread, the fringe operators began luring the unwary investors to purchase shares in claims like the North Star, Smuggler, Evening Star, Rattler and scores of others.

The town now stretched from the main townsite on the flats for almost three miles up the gulch and boasted a population of close to 500 souls. The golden years lasted for almost a decade but even they were marred by the unexpected - the 1893 diptheria epidemic which carried off

A handful of gold and silver coins recovered from the dusty remains of the basement of the old hotel.

Moffat's Place on the Hess property is the oldest surviving saloon in the area. Here the thirsty miners and freighters paused for the liquid that refreshes before continuing down the twisting road to the main townsite.

One of the inside walls of the saloon showing the squared log construction which was typical of the area.

more than a dozen of the town's inhabitants, then the failure of the many promising claims to produce as expected and finally, in 1902, a crushing blow when the pride of the town, the elegant Hotel Fairview, was levelled by a flash fire with the loss of two lives.

Perhaps it was foreshadowing, as the end was drawing near. By 1906 Fairview was on the downhill grade, the producing mines had virtually exhausted their paying ore and when values declined with depth the big mining companies began to shut their operations down. Four years later Fairview was done as a mining town and the population began dwindling rapidly. By 1919, even the physical assets began to disappear as salvage crews and fires took their continual toll. Although several attempts were made from time to time to bring some of the old properties back in production they ultimately ended in failure and for nearly half a century old Fairview lay neglected and almost forgotten.

Fairview in 1900. This photograph shows the gold camp as it appeared at its height. The buildings in the foreground are near the mouth of Reed Creek gulch. In the left background can be seen the Hotel Fairview. Nearly all of the surrounding region was arid desert country. Beyond, to the east, lay the barren flats which ultimately became the site of the town of Oliver which appeared after Fairview had disappeared.

An historic marker at Fairview.

Photograph showing detail of Fairview artifacts.

In the spring of 1969, I travelled to Fairview with an associate, Cyril W. Murray, to determine whether the region was suitable for an article. It was an unusually hot April afternoon when we arrived and on first perusal the site appeared to be nothing more than a desolate, wind swept sagebrush flat with little indication of a town or even a townsite remaining anywhere. A closer survey however, revealed the faint shadow of a town, vague outlines of old streets could be discerned and amongst the sagebrush, the rock foundations of buildings and cellar depressions could occasionally be seen. The townsite plan was gradually falling into place.

Curious, we drifted away from the gulch, down through the sagebrush towards the eastern edge of town where we knew the burned out remains of the Hotel Fairview were located. Eventually we found the spot and even though the hotel had been completely gutted two-thirds of a century before, the site was unmistakable as the hollow outline was still very evident. It was obvious that the dimensions of the "Big Teepee" had not been exaggerated. This was the place where the impressive, three story emporium had once stood and all that visibly remained was a pile of sand, rubble and a vast quantity of broken glass, the legacy of the fire. The bottle hunters had apparently been searching through the remains as a

Fifteen coins, thirteen of them silver, recovered in thirty minutes from the site. Eight were found still stuck together and had probably been in a cash box when the intense heat had adhered them. The coins, although fire blackened, were in generally good condition.

One of a number of old, derelict log structures which may still be seen in the upper reaches of the gulch.

Looking down Reed Creek Gulch. Fairview was on the rise when this photo was taken, probably in 1894 or 1895.

number of pits marked the basement area, a crude, abandoned screen indicated that a screener had also been active at one time. It appeared, however, that the greater part had not been touched since the holocaust. After considering the historical evidence and the particulars of the 1902 fire, we decided to return to screen an untouched area which we felt might yield finds of historical significance.

We returned the following week equipped with a ¼" mesh box screen and shovels. Squaring off a chosen area, we probed into the designated section. Almost immediately, a grey, flat, circular object turned up in the screen - it proved to be a fire blackened 1872 Canadian Queen Victoria quarter in excellent condition. Inside of a few minutes we had recovered a number of coins; the majority of them silver of many denominations. As we continued to sift through the sandy debris, a still glittering, circular band turned up. It was an exquisite gold ring set with five diamonds. The ring had come through the fire completely untarnished and was so bright that it appeared to have been lost only the day before, instead of almost seven decades ago. Within an hour we had recovered a number of other items including; a heavy, greyish coloured chain which later proved to be solid Sterling silver, with each link individually hallmarked; a delicate gold brooch; four nuggets of pure silver; two silver cufflinks; a slender gold ring; a small gold horseshoe, once set with semi-precious stones and a total of 27 coins which were predominantly silver. It was the first of a number of successful days of excavating at the historic hotel site.

As spring wore on we repeatedly went back to Fairview and the "Big Teepee." We meticulously screened our way through the pre-determined northern half and the finds continued; a gold-filled pocket watch with

Some of the items recovered from Fairview and the site of the "Big Teepee" Hotel. The articles ranged from coins to keys, from silver and gold jewellery to doll sets.

The Fairview jail in 1974. Since then the historic structure has been heavily vandalized. Some of the siding has been torn off, the jail bars have been removed and the door is gone.

The drugstore in Fairview. The town doctor, Dr. White is at the right side of the entrance.

Half a dozen rings, five of them gold including a diamond ring, are portrayed in the upper photograph. In the lower picture some of the other articles from the site are shown.

the hands locked at 11 minutes after 9, an English 1893 gold half sovereign in mint condition, a gold filled ring, several watch chains and more coins. After the northern part had been completed we had amassed a total of over 100 significant objects including half a hundred coins of English and American as well as Canadian origin.

Next we concentrated on the southern half of the ruins, at first working the south-eastern corner. Our initial results varied although we again recovered some fascinating articles including, four Sterling silver spoons from Birks, several silver plated spoons with "Hotel Fairview" engraved on them, two English coins, a delicate child's tea service set of china, two doll's heads and a metal toy. As we worked west, however, we again struck a rich region and found three more gold rings, a small Sterling silver chain, a gold chain, two silver lapel pins, a fine silver brooch and 18 more coins. Curiously, eight of the coins were still piled one above the other when recovered.

By now we had run through the section of the hotel which had been occupied and only the south-western corner remained. Here we diligently checker-boarded the area with a singular lack of success. It was completely barren as we had expected. Finally, after hours of fruitless digging we abandoned our project. Our timing was fortunate for in the last days of our excavating, a number of people had been attracted to the site and were beginning to dig into the ruins.

Our excavations had yielded over 150 articles and a great deal of information about both the town and the Hotel Fairview which had not been previously known. We found that many of the inhabitants of Fairview were extremely well off by the standards of the times, this was borne out by the number of gold rings and the amount of jewellery found on the site. We also discovered that silver, copper and gold coins from England were accepted at par locally. But perhaps the most important of all, we had been given an unusual opportunity to look momentarily into the past, a past when a town teemed with people, a past when a stately hotel towered over the flats and the name 'Fairview' conjured up visions of wealth and commanded respect in the Canadian West.

These are a few of the over 100 silver and gold coins recovered.

THE LOST M'LEAN

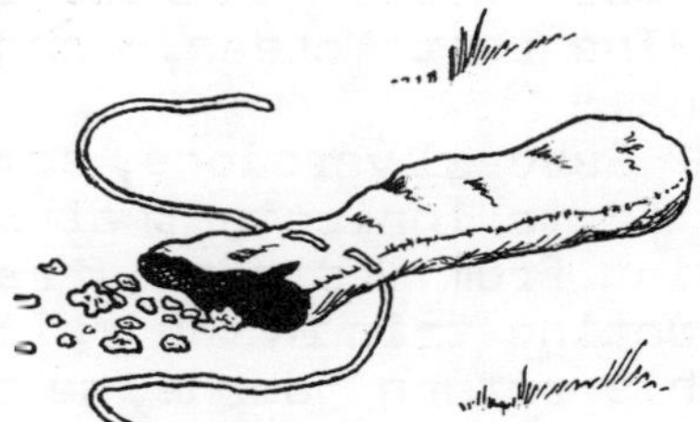

The Barlee Stretch in 1904. McLean rode this way on the way to his mine.

The particulars of this story are, in some instances, debatable but "The Lost McLean" has been included in this book because of the insistence of a number of respected local prospectors who state that the lost placer creek does, in fact, exist and because of the favourable geology and history of the area where this lost mine reportedly awaits rediscovery. For over ten years this story has gradually been pieced together. The research has sometimes been frustrating but the story which has been printed on the following pages is recorded in this book for the first time.

The origin of the story is obscure but the existence of the creek has been mooted for well over half a century. In the Okanagan it is known as "The Lost McLean," and in the Boundary Country as "The Million Dollar Creek."

Several versions, some of them questionable, exist. Generally, however, the facts indicate that sometime just prior to the turn of the century an Indian from Westbank, a reserve across Okanagan Lake from Kelowna, was on a hunting trip in the rolling hill country somewhere north-east of Kelowna. On his return journey he purportedly came across a small freshet. According to the Indian, whose surname was McLean, he camped nearby and later in the day wandered down to the creek to fetch water for his evening meal. Like many Indians of that era, he was a part-time prospector who was familiar with both placer mining and placer gold. So when he stooped to fill his billy-can with water and perceived a glint of yellow on the bed-rock, he paused to investigate. Upon close examination the glinting object turned out to be a small nugget which was rough edged and showed visible quartz. The startled Indian quickly looked at other sections of protruding bed-rock. In a short time McLean had amassed an impressive amount of placer gold, much of it with a considerable amount of white quartz adhering to the metal.

Realizing that he had stumbled across a virgin creek the Indian spent several days camped by the freshet collecting the placer gold. Finally, with food running precariously low, McLean was compelled to return to his Westbank home. His yield, gathered during his short stay on the creek, was almost six ounces. Nearly two ounces per day, an astonishing amount for such a short time, especially when he had mined without benefit of pick, shovel or gold-pan. The only equipment he had been able to use had been his hunting knife and a small frying pan.

Once back home McLean shrewdly kept the knowledge of his placer find to himself. Biding his time he waited patiently. Several months elapsed, then under the guise of another hunting trip the Indian returned to the area. It was during the middle of a typical Okanagan summer, blazing hot for weeks on end. With an unerring ability McLean was able to locate the exact spot where he had camped in the spring. The summer weather had, however, had its effect. The little stream which in the spring had been running full, had ceased running. Not even a trickle of water was visible anywhere along the bed of the stream.

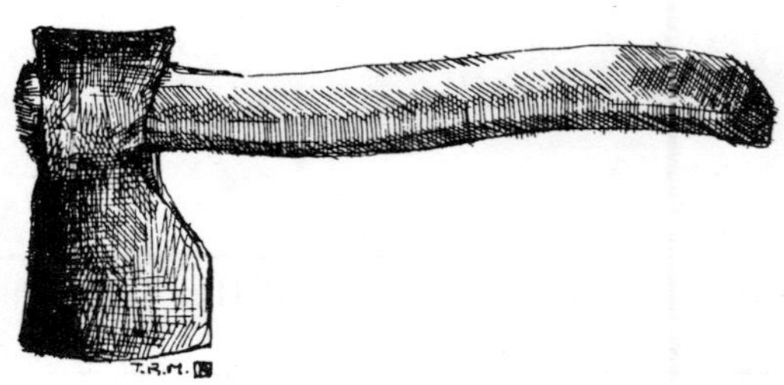

This time the Indian had come equipped for mining but with no water placer mining was impossible. He scouted along the dry creek bed and was able to recover a few nuggets and some coarse pieces of gold but it was tough going. Finally, after several hours, the Indian realized that the little creek could only be mined in the spring of the year when the water was high. By the time the sun had set the Indian had gone, back down the trail towards the distant Okanagan. This time, however, he blazed a trail which only his experienced eyes would be able to follow. His blazes were marked high on trees, well above eye level. When he returned the following spring he would be able to find the placer creek with ease.

The next spring he was back again. Equipped with gold pan, pick and shovel he quickly accumulated an impressive amount of gold. Before a week had elapsed he had filled his leather poke. Deciding that his take was sufficient he headed back to Westbank and some high living until his gold ran out. So it went each spring for almost a decade. During that span, according to local old-time prospectors, a number of individuals attempted to follow him to his secret mine, but they were never able to track the elusive Indian for more than a few miles. His usual route was along the Barlee Stretch, just east of Kelowna, then on through Rutland and on down into the canyon of Mission Creek. Once in the canyon he evidently crossed the stream and headed into the rolling hill country to the north-east. It was in this area where his trackers always lost his trail.

Finally, sometime just after the end of World War I, McLean was killed in an accident in the East Kelowna region. Unfortunately, a search of the records has failed to reveal the exact particulars of his death. With the death of the Indian the location of his mine was also lost. Since that day a number of knowledgeable prospectors have searched for the old "Lost McLean," but their efforts have been unsuccessful. The area is geologically encouraging for placer miners and the probability of undiscovered placer ground is high. There are a number of placer creeks in the general area: Mission Creek, Lambly Creek, Monashee Creek and several others. As late as 1936 new placer ground was being found. When Paul Johnson and Alf Brewer found a rich gold-bearing channel in the canyon of Harris creek near Lumby in the summer of that year, a channel which later produced coarse gold and dark hued, heavy nuggets for an experienced miner called Fosbery, it lent further credence to the long held theory that somewhere in those rolling hills north-east of Kelowna lies a forgotten creek, a creek where an old Indian named McLean found coarse gold and nuggets almost sixty years ago.

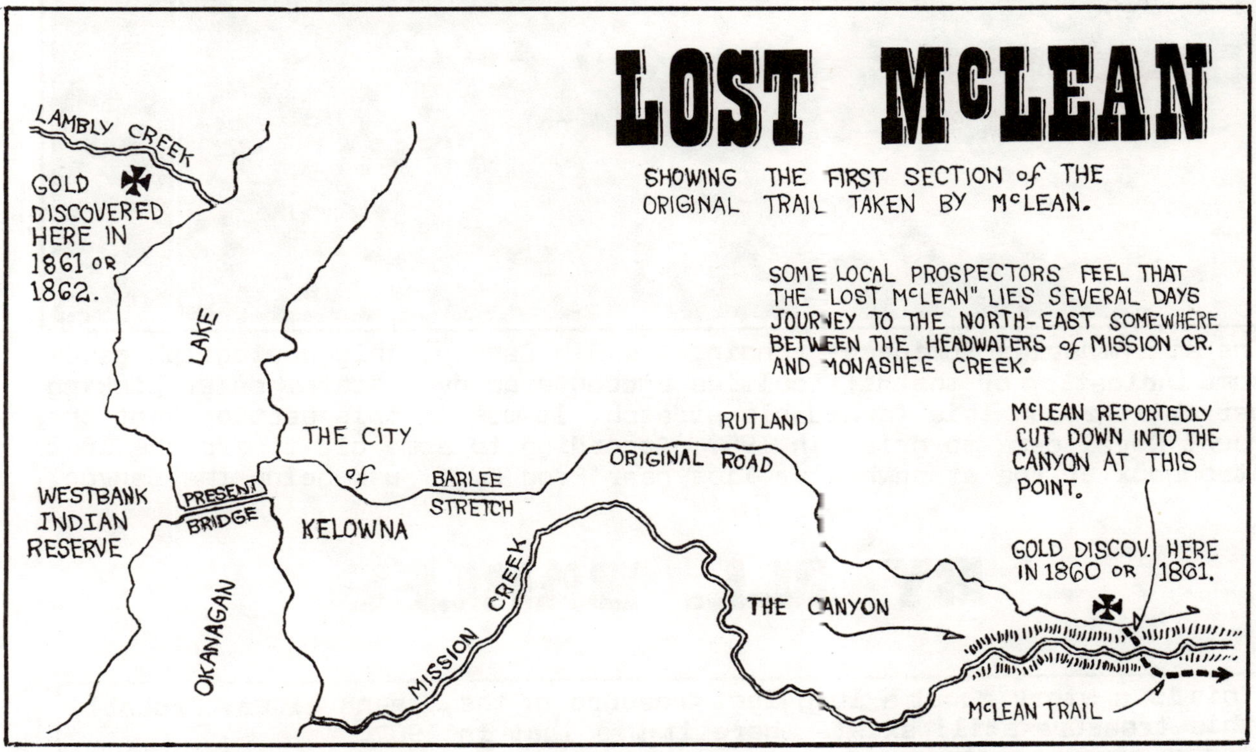

The LOST STRONGBOX of the MOUNT ROYAL

The sternwheeler "Omineca" running the Big Canyon. This photograph gives some indication of the difficulties encountered by a sternwheeler picking her way through this formidable stretch. It was in this section that the Mount Royal came to grief in 1907. According to some old-timers the lost strongbox of the sternwheeler lies near Frog Eddy just below the canyon.

BY N.L. BARLEE

This is a story about a long lost treasure of the Skeena River. Probably this treasure still exists where it was lost in 1907.

Just as the vessel entered the narrow gap the bow swung abruptly and sharply as an unexpected and unusually strong gust of wind caught her. A moment later the sternwheeler was running full tilt and out of control down the most dangerous stretch of whitewater on the river, the place feared by even the most experienced and intrepid rivermen - the Big Canyon of the Skeena.

Before Captain Stuart B. Johnson could take corrective measures, the bow, with a sickening thud, struck the looming rocks of Ring Bolt Island and held fast. The Mount Royal had bridged the narrow canyon and was broadside to the full force of the river's current.

Captain Johnson, at once realizing the seriousness of the situation, immediately ordered the gangplank lowered and several harrowing minutes later the passengers and crew were standing on Ring Bolt Island. From the security of their sanctuary they watched apprehensively as the river, running full in the narrow trough, battered the ship as it lay lodged and helpless between the rocky walls. The minutes passed and although the pounding continued, the Mount Royal, to the astonishment of the tense onlookers remained intact and looked as though it would continue to hold together indefinitely despite the incessent and incredible battering it was taking.

Surveying the scene from the vantage of the island Johnson came to a momentous and fateful decision; the sternwheeler, with just a little luck, could be saved. After a hurried consultation with the ship's officers and crew it was decided that a cable could be run through a snatch block at the stern then up to the capstan on the bow. If this operation could be completed in time the stern could then be probably be forced back and over against Ring Bolt Island and secured.

A skeleton crew consisting of four crewmen, the purser and the Chief Engineer, returned to the dangerously swaying vessel to execute this intricate manoeuvre. As the men began to carry out the various steps required to save the sternwheeler a muffled report reached the ears of the stranded onlookers on the island. Moments later the Mount Royal shuddered ominously and then quickly buckled in the middle, her guard was immediately swamped and she rolled completely over, bottom up. Seconds later she snapped in two. The king post had jumped its footing and had penetrated the bottom of the hull, dooming the vessel.

The wreckage was swept rapidly downriver and with it the unfortunate crew. Amazingly, Madigan, the Chief Engineer, trapped in the broken stern portion of the Mount Royal, rode the wreckage downstream until it finally grounded on a bar at the mouth of the canyon opposite Kitsilas village. He was eventually rescued by residents of the village after they had seen pieces of the sternwheeler floating down from the Big Canyon. He was the only survivor, the others had gone down with the stricken ship.

The passengers and crew remaining on the island were, with considerable difficulty, brought over to the mainland. As for the sternwheeler, both the vessel and its cargo were a total loss. Thus the Mount Royal, considered by many to be one of the finest ships ever to ply the Skeena, and commanded by one of the most respected and capable captains on the entire river, became a victim of the treacherous Big Canyon.

Since that disastrous day, July 6th, 1907, the Mount Royal has been the subject of much speculation. It is known that the strongbox on board the ship was lost at the time of the sinking and has never been recovered. Although there is considerable controversy concerning the contents of the safe - with estimates ranging from as high as $40,000 to as low as $700. The actual amount in cash probably approached the latter figure.

Prior to the disaster there had been numerous close scrapes in the Big Canyon because of all the turbulent waters on the Skeena River it was considered the most dangerous. But there were other places along the river designed to give even experienced captains the shakes. The names bestowed on these areas were indicative of their reputations: Hardscrabble Rapids, the Hornet's Nest, the Whirly Gig, Beaverdam and the Devil's Elbow were a few of the more treacherous places, but none of them could match the Big Canyon where the main channel was barely wide enough to accommodate a big sternwheeler and the current, especially in high water, was incredibly difficult to read.

In the inquiry following the sinking, Captain Johnson, who had a reputation for sound judgment and ability, was exonerated. It was felt that he had fallen victim to a most unusual set of circumstances: high water and an exceptionally strong gust of wind, factors which combined had resulted in the loss of the sternwheeler.

In 1907, however, business along the Skeena was still flourishing and the Hudson's Bay Company, the owner of the Mount Royal, mindful of the fact that "Old Diamond C," a businessman whose real name was Robert Cunningham (with large holdings and interests at both Port Essington and Old Hazelton) had for years competed with the Honourable Company for the

The sternwheeler "Inlander" emerging from the Big Canyon on the Skeena River. It was nearby that the Mount Royal met her end.

river trade and with his sternwheeler the "Hazelton" had given them all the competition they could handle. In the end the Bay was compelled to buy him out at considerable expense thus ending an expensive commercial rivalry which the Company did not care to repeat. With this in mind and the presence of other competitors on the Skeena the Company immediately let the contract out for a new sternwheeler. Soon the "Port Simpson," the Hudson's Bay Company's new sternwheeler was plying the river in place of the Mount Royal. In charge was the same Captain Johnson.

The sternwheeler era continued for another five years but by 1912 the completion of the Grand Trunk Pacific railway line along the Skeena had rendered river traffic obsolete. So the heyday of the sternwheeler died and soon, scrapped or transferred, the great ships disappeared from the waters of the river.

Almost seventy years have elapsed since the Mount Royal went down. In that time a legend concerning her treasure has arisen. Like many other treasure trove stories it is difficult to separate fact from fiction but there are several points which can be accepted at face value. There was a strongbox on board the Mount Royal when she sank and the strongbox did contain an undisclosed sum of money, both paper and coin, which was never recovered. There have been innumerable estimates concerning the actual amount of cash in safekeeping. Most of these estimates are grossly exaggerated and the previously mentioned sum of $700 is probably close to the actual amount lost. It must be remembered, however, that this sum was

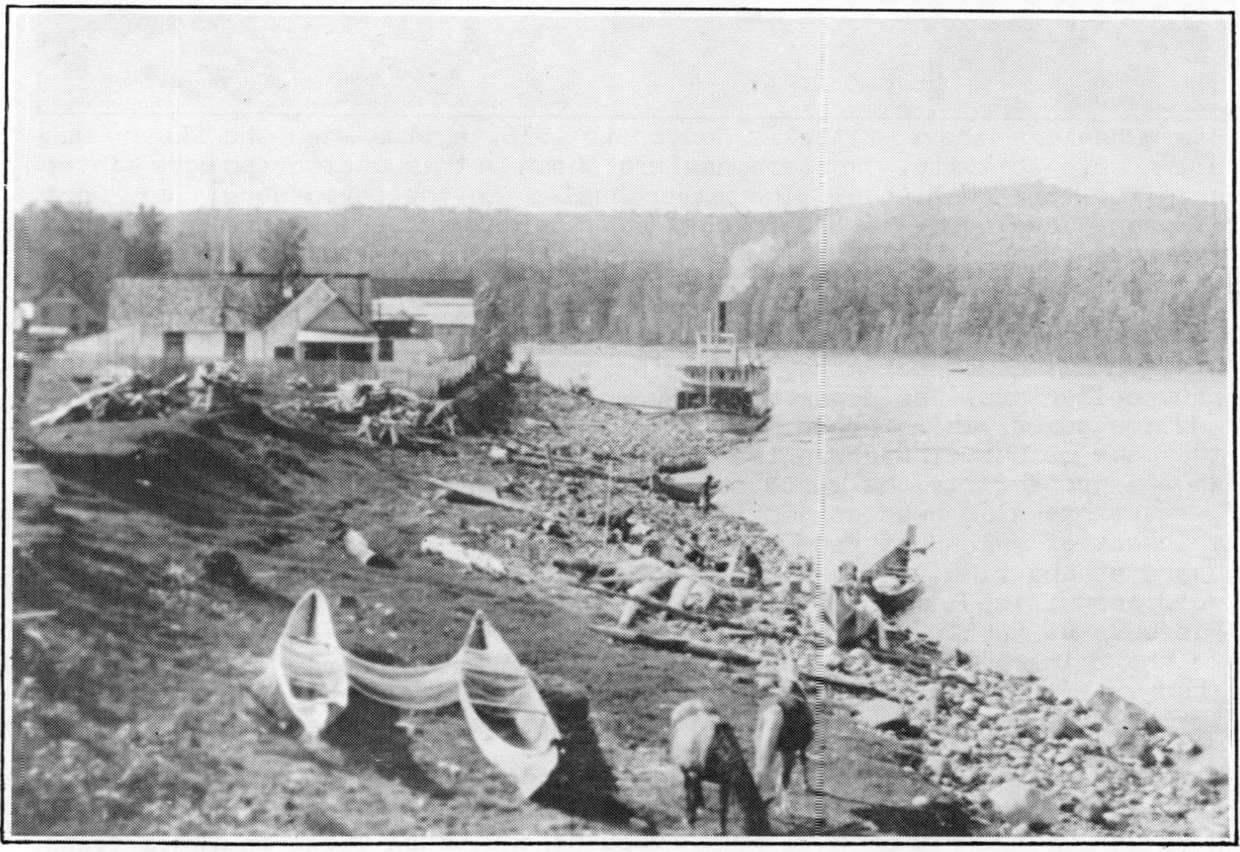

A sternwheeler tying up at one of the many river stops on the Skeena in the early days.

Sternwheelers above Kitselas Canyon in 1910, a time when the Skeena was alive with activity. The sternwheelers shown in this photograph were sister ships of the Mount Royal. The sternwheeler in the foreground, the "Port Simpson," replaced the Mount Royal and was commanded by Captain Johnson, the same officer who had been in charge of the latter at the time of the disaster.

divided between paper money and coins, both of which fetch a high premium from collectors. The paper money averages about twenty times face value and the coins, which would be mostly Victorias and Edwards, yield about the same prices. Therefore the original sum, probably about $700 would, in all probability, be worth about $14,000 and possibly more.

It has also been stated in a number of sources that there was a large shipment of gold-dust from either the Omineca country or Lorne Creek on board at the time. There is, however, no basis for this statement. The gold production from Lorne Creek in 1907 was practically nil as that creek had been virtually worked out by then and Manson and other placer streams in the Omineca were far past their peak production years. The "gold dust" theory is simply a fabrication as there is no definite evidence or records to lend credence to the claim that gold was being shipped out on the Mount Royal at the time of the disaster.

So, the River of Destiny, as the Skeena is sometimes known, still guards her secret which has fascinated people for decades and has induced a number of them to embark on infrequent and usually ill-equipped tries to recover the hoard, attempts which have all failed.

A Skeena River sternwheeler.

Taking wood on board. Everyone helped.

But that memorable river hasn't changed much: the boats are gone but the ring bolts are still anchored to the perpendicular walls of Kitselas Canyon and encircle Ring Bolt Island - and just below the mouth of that canyon, somewhere on the bottom of the river, perhaps in the depths of Frog Eddy and within sight of the long abandoned old village of Kitselas, lies the rusted strongbox of the ill-fated sternwheeler once known as the Mount Royal.

The Mount Royal on her maiden voyage before her tragic end in 1907.

Some of the photographs used in the above story were provided by Mr. Don Steele and much helpful information was provided by the works of the late Wiggs O'Neill of Terrace. After the original story was published in "Canada West Magazine" in the spring, 1972 issue, additional information was provided by Eric R. Thomson of Hopkins Landing, B. C. whose father, James Thomson, had been the HBCo. manager at Victoria from 1896 to 1911. Mr. Thomson stated that his father told him that the strongbox on board the Mount Royal at the time of the disaster contained "about $700."

The photographs in this article have been provided by D. Steele and the British Columbia Provincial Archives. The original map by the late Wiggs O'Neill of Terrace was also a valuable source.

The 10,000 LOST COINS

The main street of Sandon as it looked in the early 1950s. Black's store, the post office, the Burns building, Hunter building, the Virginia Block. The Palace, Reco and Sandon hotels are also visible to the right.

BY N.L. BARLEE

Inside a four mile radius in the Slocan Country four treasures have been recovered. This story is about yet another hoard, thousands of coins which were lost decades ago in a remote ghost town called Sandon. These coins, most of them silver and some gold, have a high numismatic and historic value and have been the focus of treasure hunters for almost two decades. This is the first time the details of this story have been disclosed.

There are probably 10,000 coins lying somewhere along the bed of the creek. The value of these coins; various denominations of gold, silver, nickel and copper, is estimated at close to $20,000.

It's a strange story involving a disastrous fire, a main street which had been built over a rampaging creek and the failure of treasure hunters to locate the hoard although they know its general location.

The concentration of coins almost certainly lies along a stretch of the creek measuring no more than one-third of a mile in length. The coins undoubtedly lie at a relatively shallow depth, probably not more than three feet and possibly less.

Although several hundred of these coins have been recovered in the past few years the rest remain where they were washed by the current of Carpenter Creek when it flooded in 1955.

The story begins in the 1890s. In that era the "Silvery Slocan" was one of the illustrious mining regions in the province. And in the heart of the silver belt stood the city of Sandon, a mining town which, in its day, had few equals. It was a high riding camp, especially in the early years of that decade. Its single crooked main street was crowded with all types of businesses and its boardwalks jammed with people. For seven years

The back bar of the Reco Hotel in the early 1960s. Although most of the original hotel bars from this town were destroyed or taken across the line into the United States. This bar is still in Canada.

The Sandon Hotel as it looked in the early 1950s before the flood. (Copyright Canada West)

Sandon was the commercial centre of the Slocan region, challenged only briefly by Kaslo to the east and upstart New Denver to the west.

And for seven years the mines of the Slocan poured out their wealth, a share of which found its way into the many saloons of Sandon. On May 3, 1900, however, disaster struck when a fire which started in the opera house quickly swept out of control and roared down the main street. In its wake it left a city of ruins. In the business district only a single structure remained standing. Sandon, the pride of the Slocan, had virtually disappeared.

Within weeks the city's business district had begun to rebuild. But the new city was far different from the old one. In the 1890s the main street had been situated on the north side of Carpenter Creek. This time, however, it was decided that more room was needed so the creek was flumed and boarded over. The boarded over creek became the new main street and newly constructed businesses were thrown up on both sides.

Unfortunately, the post-1900 Sandon never equalled its illustrious predecessor. The boom years had passed and the mines were beginning to run out of rich ore. The fate of the mines also decided the fate of the town. Although there were brief respites when new ore deposits were found by the late 1930s Sandon had slipped badly and was virtually a ghost town. Most of the once flourishing hotels and stores along the main street were closed and boarded up. A few, like Johnny Harris's elegant Reco Hotel and the Virginia, remained open but it was a futile gesture - the town of old was dead.

For a few years during the early 1940s when the Japanese-Canadians from the Pacific coast were transferred to the interior, Sandon, like a number of other old towns in the district, came alive once again. But it was a short-lived respite. In the spring of 1955 Carpenter Creek went on

Carpenter Creek roaring out of the flume under the main street of Sandon in the 1950s just before the flood.

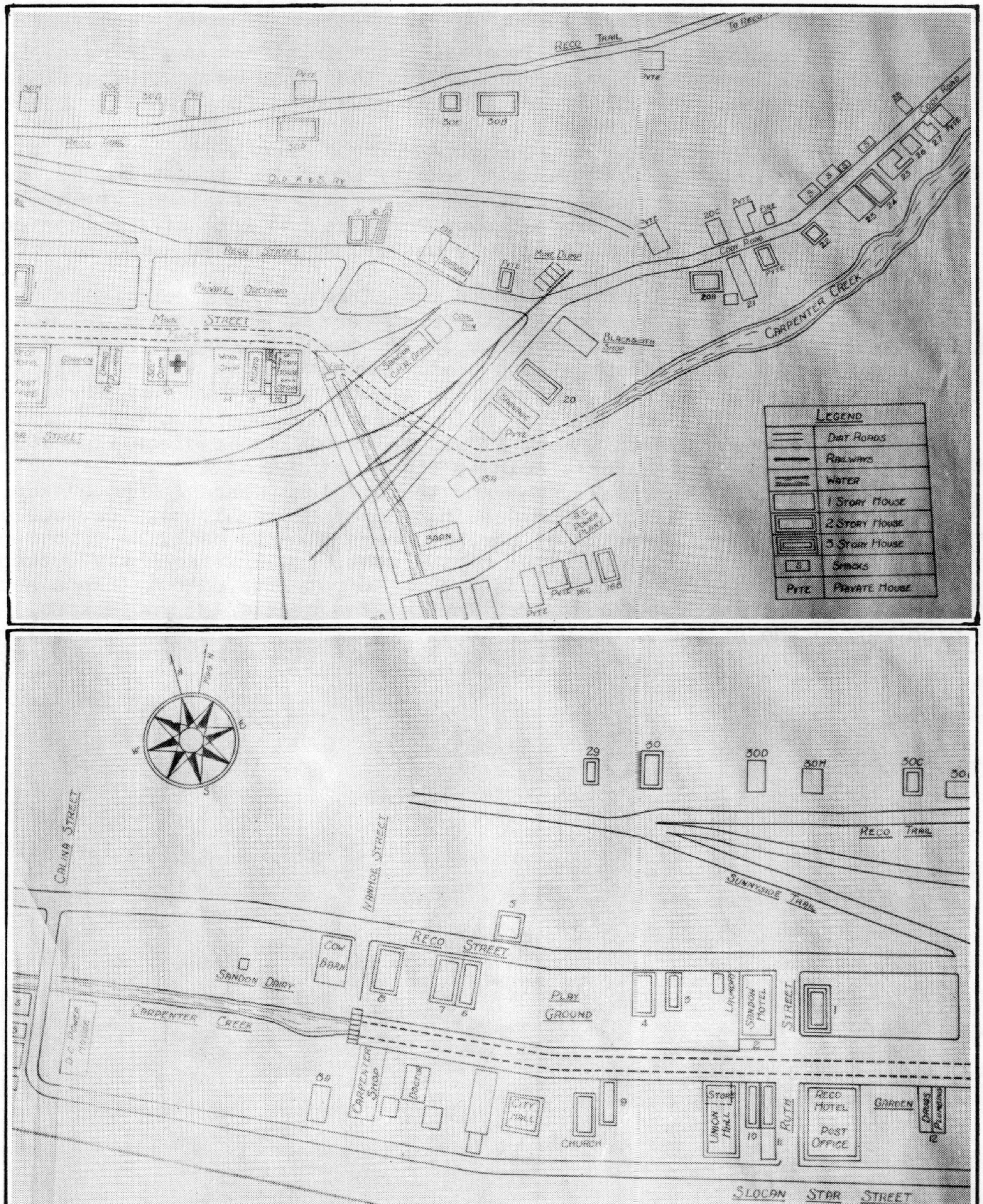

The Roy Matsui map of Sandon, 1942. (Copyright by Canada West Magazine)

the rampage, tore the flume apart, wrecked the boardwalk and swept down the main street.

The destruction it left was immense. The main street was in ruins, a number of buildings had been carried away by the flood waters, others had been undermined and tilted. It was a horrendous blow for Sandon and one from which it never recovered.

Even at this late date the town could have been saved but when the local salvagers, primarily from New Denver, moved in, it was doomed. By the 1960s little of the once impressive main street remained. The swath cut by Carpenter Creek had widened over the years and most of the land on the north side of the gulch had slowly washed away and had been carried downstream.

This was the condition of Sandon when I returned to the town in the late 1960s after being absent from the area for nearly twenty years. I had purchased some buildings and property in the town, hoping to eventually restore some of the grandeur of the past. One afternoon while absently walking along the last remaining section of boardwalk that had survived the 1955 flood I picked up a silver coin. Carefully rubbing off the grime I examined it. To my astonishment it was an exquisite 5¢ piece - an 1896 Victoria. It was a collector's coin in almost mint condition.

I gazed at the boardwalk. Many of the original boards were broken, others were loose. The structure was in general disrepair and obviously beyond preservation. Memories of earlier years flooded back. As a boy in Rossland, the famous "Golden City" of the west, I had frequently ducked under the old boardwalk on Columbia Avenue to retrieve coins, tokens and other valuables which had been lost through the cracks in the boards. A high percentage of the coins had also been Victorias. The historic value of the coins had not concerned me then but my attitude had changed since

The upper gulch in Sandon after the 1955 flood.

The Miners Union Block in the early 1950s. This photograph was taken just before the flood of 1955 and shows part of the main street and the boardwalk under which flowed Carpenter Creek.

those distant days.

Sandon, like most of the old western towns, had once had a boardwalk the full length of its main street. In fact it had had two sets of boardwalks, the first on Reco Street when it had been the main thoroughfare in the 1890s, and the second on Main Street in the "new" town after the fire of 1900.

The 1955 flood had, however, complicated matters considerably. Only a small section of the Main Street boardwalk had survived, about sixty or seventy feet in all. And of the original boardwalk on Reco Street nothing remained - it had been completely gutted by the 1900 fire.

Innumerable coins, I reasoned, must have been lost through the cracks of both sets of boardwalks in those years when the city was flourishing. I decided to check out this possibility. In front of the Virginia Block (which is owned by Canada West Publications) I carefully screened some of the dirt from under the old boardwalk. Four hours later I was rewarded with nine coins, most of them silver, and ranging in dates from the early 1890s to the late 1940s.

Eventually I screened nearly fifty feet of the boardwalk and found nearly 100 coins, many of them in almost mint condition. Although I had checked only the last 50 feet of the boardwalk, I had recovered two coins

The city of Sandon sixty days after the fire of 1900.

A few of the coins recovered from under the last section of boardwalk in Sandon. The coins date from the 1880s through to the early 1950s.

per running foot of boardwalk and assumed that at least fifty percent of the coins had been washed downstream during the flood.

I studied the map of the town and quickly came to the conclusion that there had been a total of just over 4,200 feet of boardwalks in the two Sandons and that almost all of this total had been destroyed in the fire of 1900 or during the flood of 1955. If the boardwalk averaged only three coins per running foot, and this was, I felt, a conservative estimate, then more than 12,000 coins had been washed downstream by the flood. I knew from research that, apart from the 100 I had found, few others had been recovered. The remainder, numbering at least 10,000 lay downstream. And they are still there, almost $20,000 worth of old coins, somewhere along the bed of Carpenter Creek, near the old town of Sandon.

Please note - Much of the old town of Sandon is privately owned by local and outside interests. Individuals intent on recovering treasure in this area could be prosecuted.

View of the upper part of Sandon in the late 1890s.

The LOST CHRISTIE LEAD

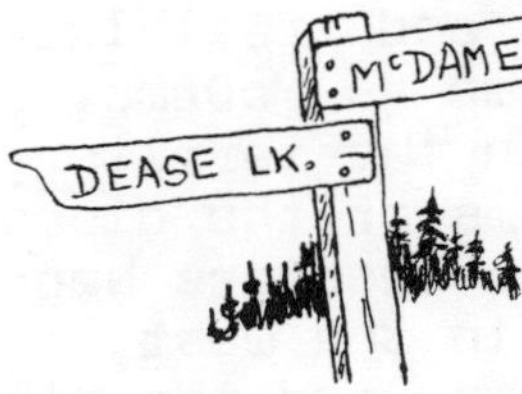

A prospector leaving Telegraph Creek in 1898. When this photograph was taken the Cassiar district had declined in importance as a gold producer.

It was once the most defined paystreak in the province and the gold from it so heavy that even veteran miners were astounded at its coarseness. In the 1870s it was known far and wide as "The Christie Lead" and it still is – a run of gold that was nearly one hundred feet wide and continuous for almost a mile until it abruptly and unexpectedly vanished. But there are placer miners in the Cassiar Country who swear that somewhere along McDame Creek the old lead exists yet – awaiting rediscovery.

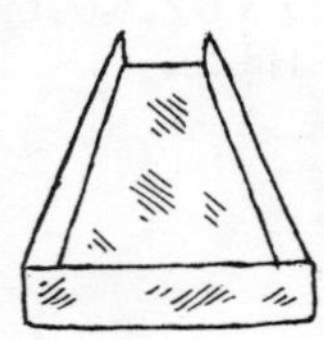

The gold was amongst the coarsest ever found in the province with nuggets weighing anywhere from an ounce to two-thirds of a pound. The original discovery was made in the spring of 1876, almost a century ago.

In 1873, after the great Cariboo gold rush had subsided, the name "Cassiar" quickly came to the fore when Thibert and Dease creeks were discovered to be gold bearing. The following year the redoubtable black prospector and trailblazer Henry McDame discovered a remote stream which became known as McDame, a creek which ultimately became the premier gold producer in the territory.

As news of the discoveries filtered through to other regions of the province prospectors began to stream into the area. Like placer strikes elsewhere in the west, the shallow ground was worked first and later the deeper ground and the higher benches. On McDame creek it was soon found that the richest section of the creek was upstream from the junction of McDame and a tributary called 1st Fork, west of a motley collection of log cabins and stores generally known as "Centreville," a miner's camp which served as the trading centre for the miners strung out along the stream.

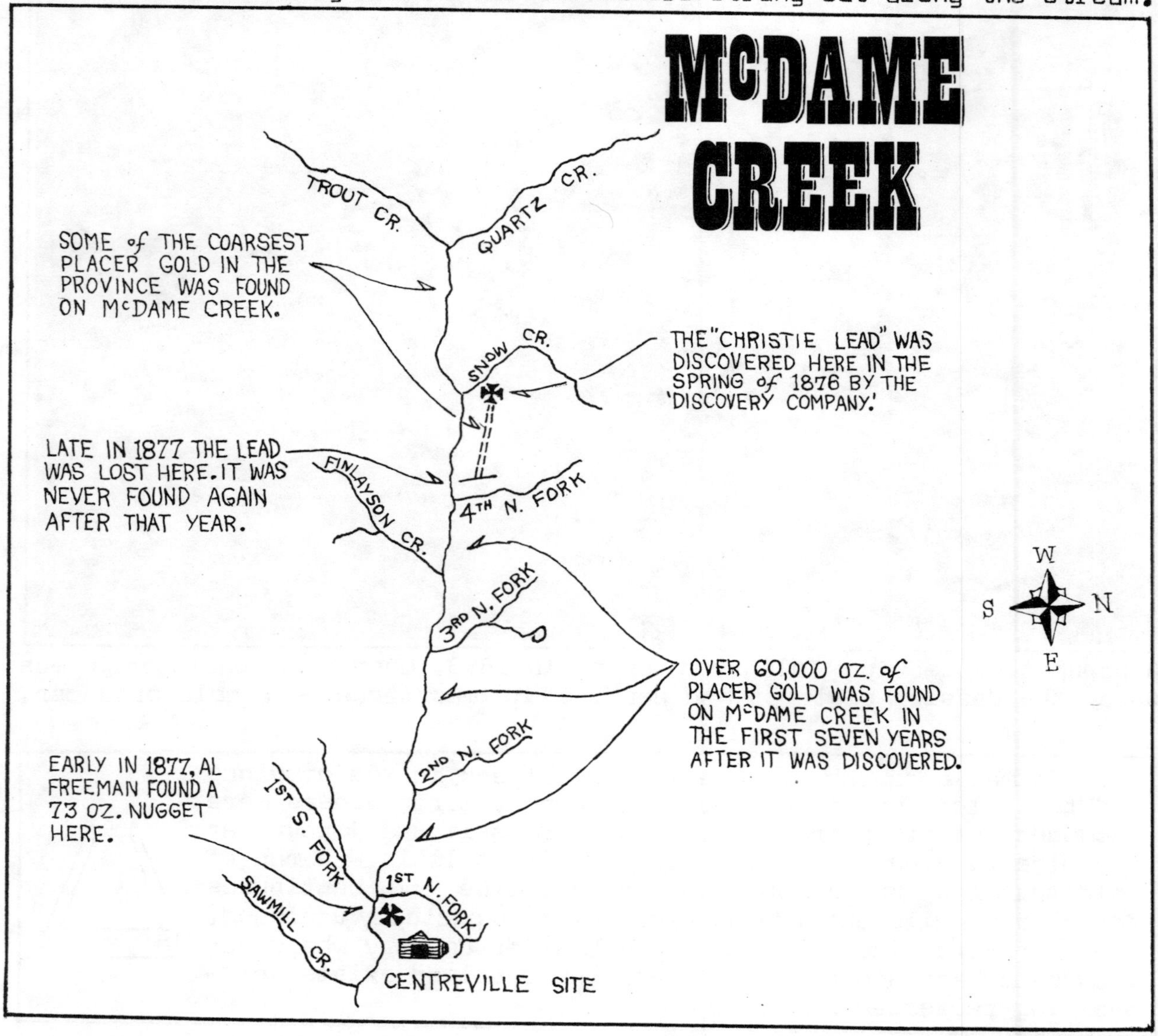

By 1876, however, the gold on Dease and Thibert creeks was beginning to run out. McDame creek, fortunately, was still productive as companies like the Charity, Black Jack, Caledonia and a score more were making huge profits. Early in that season, however, an event occurred which was to electrify the camp. The Discovery Company which consisted of Fred Black, Rufus Sylvester, Vital LaForce and a miner called Christie, all old Cariboo hands, were at work on Snow Creek, a tributary of McDame creek several miles upstream from Centreville.

While working their claim near the mouth of Snow creek canyon they blasted a boulder about forty feet from the mouth. Christie, who had been in charge of the work, casually panned the gravel which had been under the rock to determine whether or not it was gold bearing. To his astonishment he discovered that the gravel was shot through with coarse gold and nuggets. It was the start of the famous "Christie Lead."

From the gravel where the boulder had stood 73 ounces of coarse gold was recovered. The lead, which was located on the bottom side of the first bench, seemed to head east toward the 4th North Fork. The four miners immediately began following the run. The lead, nearly a hundred feet in width, soon proved to be consistently rich. The yield, steadily averaging between 125 and 375 ounces was returning between $2,000.00 and $6,000.00 weekly. All through that summer the four partners stayed with the run as it continued to give up nuggets and coarse gold. The fame of the discovery spread up and down the creek and along its numerous tributaries. Soon Centreville buzzed with activity as scores of the 700 miners on McDame began probing into the benches in search of other runs or the extension of the Christie Lead.

In 1876, for the first time, McDame creek became the premier placer gold producing stream in the Cassiar, surpassing the output of both Dease and Thibert creeks. The gold from the Christie Lead contributed a hefty portion of the total of $163,700.00 for the creek for the season.

During the winter the Discovery Company was able to obtain the bench claims which were in line with the original location. This was a shrewd move on the part of the four partners because in the spring of 1877 they were able to continue following the rich lead farther into the bench by alternately drifting and hydraulicking. And the gold, heavy and nuggety, continued to pour out of the paystreak. The weeks turned into months and still the lead continued eastward toward the 4th North Fork. The lead was nearly 100 feet wide and almost 20 feet deep and appeared that it would continue almost indefinitely. But towards the end of the season the run suddenly petered out - the "Christie Lead" had stopped.

Convinced that it was only a temporary setback the four miners began drifting and sinking shafts in the area to try to pick it up again. By the end of the season, however, despite countless efforts they had failed in their attempts to find it again. For some inexplicable reason the famous lead had simply disappeared.

For several more years other individuals and companies attempted to re-locate the old lead. Tunnels extending over 1500 feet were driven in anticipation of picking it up but none was successful. By the end of the 1877 season all of the exploring companies had been bankrupted. By 1880, nearly half of the 245 miners still on the creek were Chinese.

In the decades since the "Christie Lead" disappeared other attempts

Ground sluicing on the Zimich lease on McDame creek.

have been made to trace the original run but these, like the others of a century ago, have also ended in failure.

Today there are still placer prospectors in the Cassiar country who flatly insist that sometime someone will re-locate the lost lead and when he does the yield will parallel those days of a century ago, those days when the phrase "Christie Lead" meant coarse gold and nuggets and McDame creek was more than just a name.

A grave above McDame creek. The inscription on the wooden marker is still as sharp as it was sixty years ago.

The information used in this article has been obtained from the Annual Reports of the Minister of Mines, from the Archives of British Columbia and from various individuals in the Cassiar area including Jack Harley and George Zimich who still have placer leases on historic old McDame creek.

THE GOLD POOL of the SIMILKAMEEN

The writer examining two tobacco tins he found in the old Bowman cabin on the Knob Hill claim.

BY N.L. BARLEE

This is a story about a "gold pool." On a remote part of the Similkameen river a peculiar geological oddity has resulted in the formation of a series of rocky underwater riffles past which no placer gold passes.

The old Knob Hill mineral claim lay to the west of the river. On the eastern side the tunnel of an unnamed copper mine stood out against the sheer canyon wall. I was in one of the most remote parts of the Similkameen country, in an area which only a handful of people knew even existed.

On the way in I had come across a long abandoned and tumbled down log cabin. The roof was gone and although little else remained except the sagging walls, I was curious because it was obvious that few people had ever passed that way.

Stepping through the doorway I surveyed the remains of the cabin. Decayed wood lay intermingled with broken roof beams. In one corner a hand-made wooden shelf, tilted crazily, had sagged to the floor. I picked my way through the debris, lifted the shelf and peered underneath. Two old tobacco tins lay in the corner. One, an 'Old Chum,' was about half a century old and relatively common but the other was a real prize. It was a 'Kelowna Tobacco' tin in almost perfect condition. The label read 'coarse cut.' Anybody who had ever smoked 'Kelowna' would undoubtedly have agreed that all of that particular brand of tobacco was very coarse indeed. But 'coarse' or not it was a real find. I was aware that there were probably no more than twenty or twenty-five Kelowna Tobacco tins in existence. Both tins, fortunately, had been protected from the elements when the shelf had fallen so many years before. Caching my find to await my return journey I continued on down the trail.

A few minutes later I was standing on the rim of the canyon. Below

The two tobacco tins lay in a corner of the abandoned cabin under an old wooden shelf.

me lay the Similkameen river. Although it was fall and the river was low the water cascaded through the narrow gorge with surprising velocity.

Years before I had heard about a small gravel bench in the canyon which had annually yielded interesting amounts of placer gold after high water. Aware that the gold had been washed downriver by the spring run-off I had hiked into the area to check it out and to ascertain whether or not the story was worth following up.

An overgrown trail led down the side of the canyon to the river below and I cautiously made my way down to the canyon floor. Bedrock was showing all along the western side and upriver, just around a sharp pinnacle of rock, a gravel bench hugged the canyon wall. I glanced upriver. The river swung sharply into the bench - it had to be the place.

Edging my way along the wall I stepped onto the bench. A few minutes later I was test panning the gravel. Several minutes passed before I had my answer - fifteen coarse colours lay glittering on the bottom of the pan. I gauged the result, it was almost a dollar's worth of gold. It was a good pan.

For most of the afternoon I worked along the narrow bench. I tested the surface gravel and the gravel lying on bedrock. The bench was only about twenty-five feet in length but every pan, without exception, had yielded coarse gold, anywhere from 25¢ worth to a dollar. It was rich, no doubt about it. But where had the gold come from?

Later that day I made my way upriver, testing the gravels as I went.

The tunnel had been driven into the precipitous wall of the canyon. The old workings were caved and the original trail to the mine had vanished.

Several hours later I was back on the bench, no more knowledgeable than I had been earlier. I had tested numerous places on both sides of the river but had been unable to determine where the gold had originated. When I left later on that day I had a nice showing of gold and its origin had whetted my interest.

Intrigued by the area and its possibilities I returned each fall for several seasons and was always rewarded with more gold. The high water in the spring always restocked the little gravel bench with fresh gold. And gradually I began to obtain a clearer picture of the canyon area. Then one autumn day something even more fascinating caught my attention. Although I had never noticed it before, a distinct fault extended from one side of the river to the other. Barely perceptible at river level the dark break line was easily seen from the rim above when the light was at a certain angle.

Interested, I made my way downriver, past the deep pool created by the fault. Beyond the lower limit of the canyon I panned the gravel close to the river's edge. No colours! I forded the river and tried the opposite side. Again no colours. There was placer gold in the gravel upstream from the pool but no gold downstream. It was apparent that the gold which was being continually washed downriver, especially in the spring, was being deposited in the pool.

Ordinarily experienced placer men will ignore potholes or pools where novices are sure gold collects. The reasons are quite logical: the water turbulence in these places is so great that the current simply washes the gold on downstream, or the rocks carried downriver in the spring act like a gigantic ball mill which grinds the gold to a fine powder, this powder, in turn, is then blown out of the pothole or pool and carried away by the current. This pool, however, was somewhat out of the ordinary for several reasons: there was little perceptible turbulence at the fault line, and secondly, the pool was at least twenty feet in depth, deep enough to allow any gold being carried downstream to settle on the bottom of the pool. In other words, the pool seemed to be a huge but effective river riffle.

The prospects were intriguing. It seemed plausible that the pool was a vast depository of gold but there remained some unanswered questions. How much placer gold was actually lying on the bottom of the pool? Would it be worthwhile going after? I weighed the chances and decided to check it out.

Several weeks later I called in on the Gold Commissioner in nearby Princeton. An examination of the Placer Claims map showed that the placer mining lease on Lot 709, the old Knob Hill claim, had reverted to the crown. I immediately picked up some claims tags and headed for the claim. An hour later I was driving in my claim posts and by the end of the week I had applied for the lease.

During the fall I carefully checked the lease. On the upper part of the claim, near the mouth of Friday Creek, a short-lived hydraulic operation had once been tried. On the lower end of the claim the old-timers had ground-sluiced a large gully. Between the gully and Friday Creek I had recovered a considerable amount of gold. Some of the gold, interestingly enough, had quicksilver on it, a sign that it had once escaped through the sluice-box of a prospector of long ago. The gold had to be coming from somewhere on the upper section of the claim. There were a number of possibilities. It could have come down Friday Creek or from one or the other side of the canyon and probably the west side where there was still evidence that the old time prospectors had once worked an ancient channel high above the canyon.

So far I haven't found the origin of the gold and the gold pool is still there in that forbidding section of the river. The Similkameen has been flowing over the pool for decades depositing its annual tithe of gold dust and nuggets. Someday I'll have some divers probe the depths of the pool - and then the riddle of the canyon will probably be solved.

The original post of the Knob Hill claim stands yet above the river.

The CASCADE TREASURE

First Avenue in Cascade City, probably in the summer of 1899.

BY N.L. BARLEE

This is an account about a treasure, in all likelihood a treasure which still remains where it was lost when a fire raged through a frontier town over three-quarters of a century ago. A cache of silver coins, and perhaps a few gold coins as well, still lie in the basement ruins of a hotel in a town which was once known as Cascade City.

On September 30, 1899, a flash fire ripped through the town of Cascade in the eastern Boundary Country of British Columbia. Perhaps it was not totally unexpected as the town had been experiencing a severe recession and business decline since the year had begun.

The holocaust, almost certainly of incendiary origin, and wildly out of control had roared along First Avenue. Finally, after ten businesses had been destroyed, the fire was brought under control. But First Avenue, the main business street of the town, was a shambles.

The CASCADE RECORD faithfully reported the details of the fire on its front page. The usual particulars were included but one paragraph was of more than passing interest. Part way down the front page a column about the fire stated:

> "The Chinese cook at the Grand Central carefully carried a ham out to safety, and left $70 in money to be burned up in his room."

Ordinarily a report of this nature would not attract much attention but there were several points worth noting. First, the Chinese invariably turned their wages into hard cash, that is, gold or silver currency and preferably the latter. They were suspicious of paper money and rightly so because they had been victimized on a number of occasions by unscrupulous whites. Secondly, it a common belief that fires melt down silver and gold coins. This is generally an erroneous assumption. Silver is occasionally affected but only when the fire is of long duration. Gold coins, on the other hand, are rarely affected because the peculiar properties of that

Looking down Main Street in Cascade sometime around 1909.

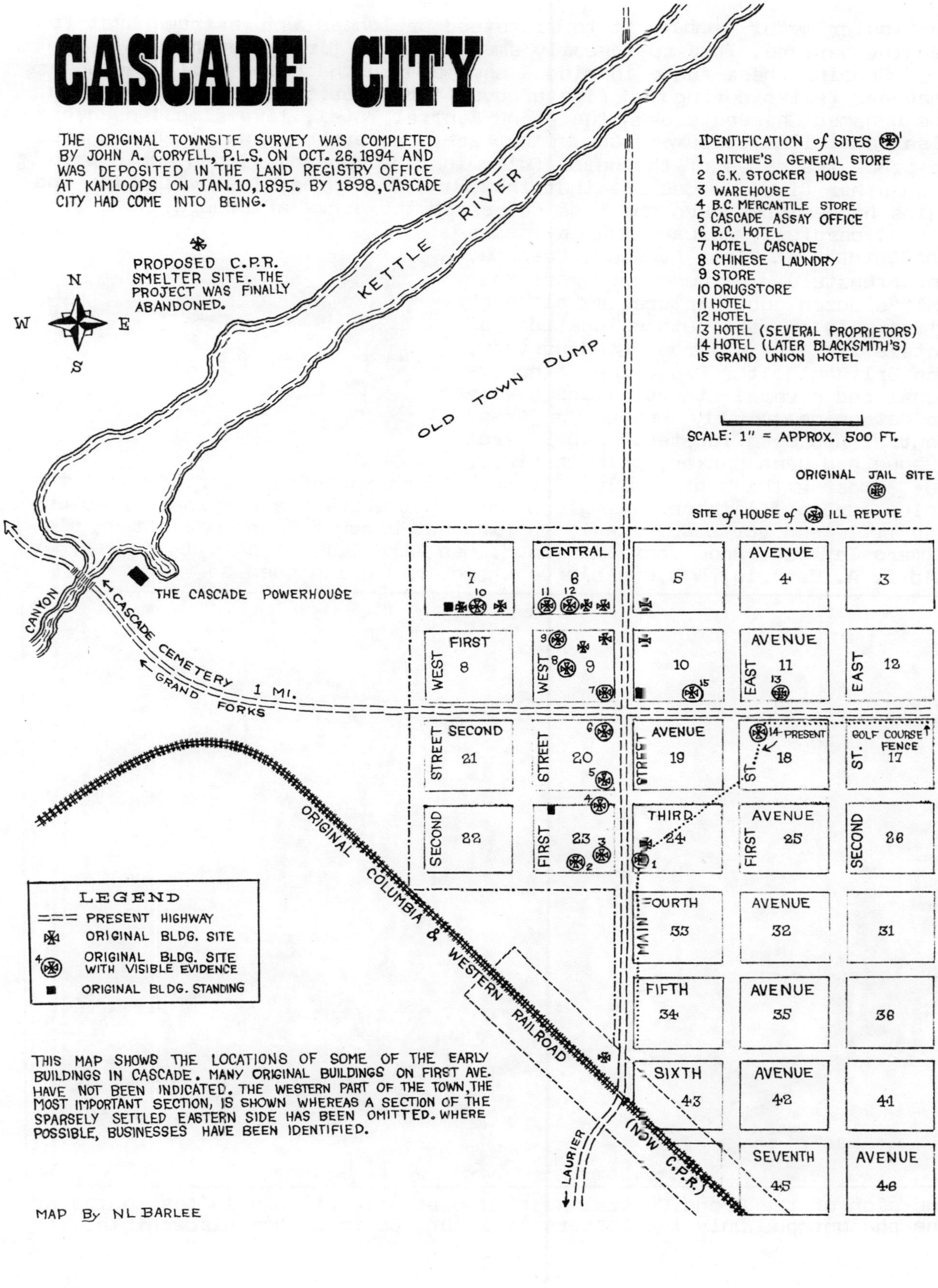
CASCADE CITY
THE ORIGINAL TOWNSITE SURVEY WAS COMPLETED BY JOHN A. CORYELL, P.L.S. ON OCT. 26, 1894 AND WAS DEPOSITED IN THE LAND REGISTRY OFFICE AT KAMLOOPS ON JAN. 10, 1895. BY 1898, CASCADE CITY HAD COME INTO BEING.
PROPOSED C.P.R. SMELTER SITE. THE PROJECT WAS FINALLY ABANDONED.
N
W
E
S
KETTLE RIVER
OLD TOWN DUMP
IDENTIFICATION of SITES
1 RITCHIE'S GENERAL STORE
2 G.K. STOCKER HOUSE
3 WAREHOUSE
4 B.C. MERCANTILE STORE
5 CASCADE ASSAY OFFICE
6 B.C. HOTEL
7 HOTEL CASCADE
8 CHINESE LAUNDRY
9 STORE
10 DRUGSTORE
11 HOTEL
12 HOTEL
13 HOTEL (SEVERAL PROPRIETORS)
14 HOTEL (LATER BLACKSMITH'S)
15 GRAND UNION HOTEL
SCALE: 1" = APPROX. 500 FT.
ORIGINAL JAIL SITE
SITE of HOUSE of ILL REPUTE
THE CASCADE POWERHOUSE
CANYON
CASCADE CEMETERY
GRAND FORKS 1 MI.
CENTRAL AVENUE
FIRST AVENUE
SECOND AVENUE
THIRD AVENUE
FOURTH AVENUE
FIFTH AVENUE
SIXTH AVENUE
SEVENTH AVENUE
WEST STREET
SECOND
WEST STREET
FIRST
MAIN STREET
EAST ST.
FIRST
EAST ST.
SECOND
PRESENT
GOLF COURSE FENCE
ORIGINAL COLUMBIA & WESTERN RAILROAD (NOW C.P.R.)
LAURIER
LEGEND
PRESENT HIGHWAY
ORIGINAL BLDG. SITE
ORIGINAL BLDG. SITE WITH VISIBLE EVIDENCE
ORIGINAL BLDG. STANDING
THIS MAP SHOWS THE LOCATIONS OF SOME OF THE EARLY BUILDINGS IN CASCADE. MANY ORIGINAL BUILDINGS ON FIRST AVE. HAVE NOT BEEN INDICATED. THE WESTERN PART OF THE TOWN, THE MOST IMPORTANT SECTION, IS SHOWN WHEREAS A SECTION OF THE SPARSELY SETTLED EASTERN SIDE HAS BEEN OMITTED. WHERE POSSIBLE, BUSINESSES HAVE BEEN IDENTIFIED.
MAP By NL BARLEE

particular metal enables it to withstand prolonged and extreme heat for lengthy periods. They customarily emerge from a fire untouched.

Bearing these facts in mind I assumed that the coins still lay where they had fallen during that fire of over three-quarters of a century ago. The unnamed Chinese cook at the Grand Central Hotel, like almost everyone else in similar circumstances in that era, had probably not even bothered to probe the ashes of the hotel following the tragedy. I had ample proof from other fires in Cascade that the victims who had lost valuables and coins had seldom taken the time to search the ashes afterward.

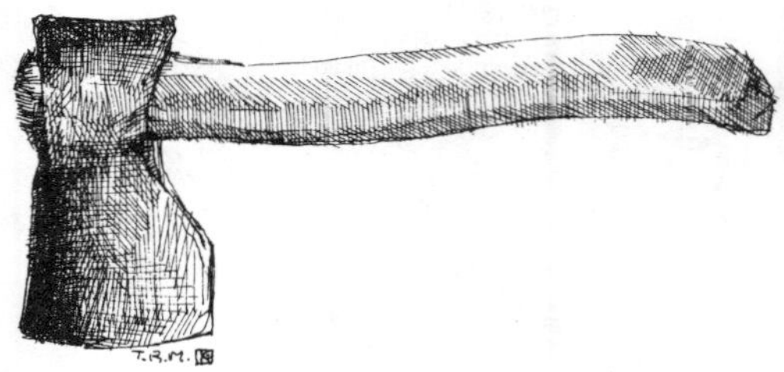

I carefully perused the existing photographs of First Avenue in Cascade. Unfortunately, there were not more than half a dozen good pictures and although I was able to make out the locations of hotels like the Railroad Headquarters, the Britannia, the Club, the International and several others, I was unable to determine exactly where the Grand Central had been located. In 1899 First Avenue had been packed, on both sides, for almost a third of a mile. Although I was reasonably certain that the coins were still there I simply didn't know where the original structure had been and nor was I likely to find out because there weren't any old-timers left who knew where it stood. Men like R. G. Ritchie, G. K. Stocker and J. A. Bertois, who would have known, had all passed on.

The sign of the Grand Central Hotel is partially visible in the centre of the photograph. Only the letters 'ral' of 'Central' are discernible.

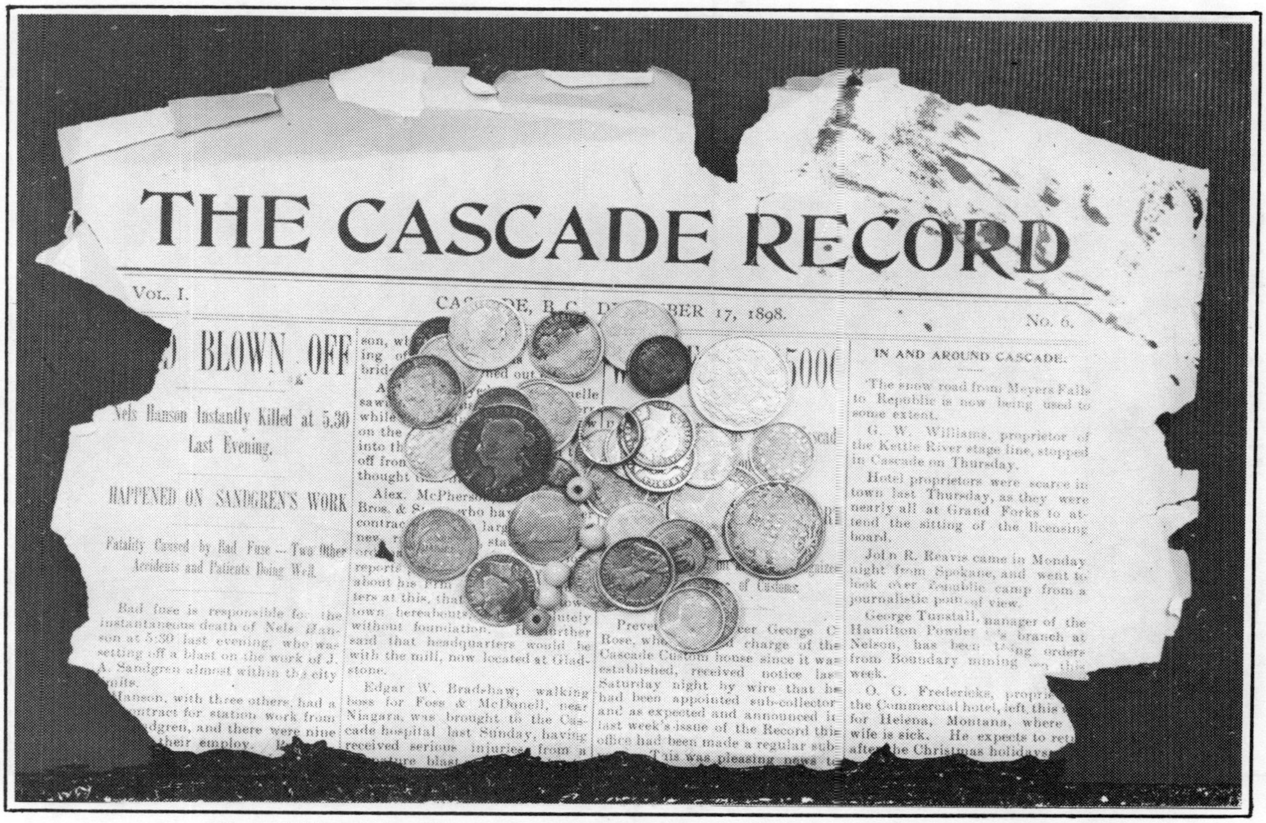

Old silver coins from Cascade recovered by the writer.

Looking from the basement sites on Main Street toward First Avenue.

The power dam near Cascade. The gorge is visible under the trestle.

To further complicate matters I assumed that the Chinese cook had lived at the back of the hotel. He would have undoubtedly hidden the coins in his living quarters, but I had no idea of the length of the hotel, it could have been twenty feet or sixty feet.

The coins would have been either American gold coins or Canadian or American silver coins dated 1899 or earlier. If they were Canadian silver coins they would be prized Victorians, probably of various denominations and in varying condition. A conservative estimate would probably place their numismatic value at twenty times their face value. The original $70 would be worth about $1,400 today.

Several other factors made the search interesting. First Avenue was not the same as it had been in 1899. Most of the basement excavations had been filled in over the years and treasure hunters, mainly Americans, had combed the facing part of the avenue. They had done well, so well in fact, that I had recovered only two coins after a careful search of the front part of the old abandoned avenue. My calculations indicated that they had probably found anywhere from 400 to 800 historic coins along the street. Later research, however, showed that they had concentrated only on the fronting side of the avenue and had not bothered with the back parts of any of the numerous sites along the street.

Although I spent several days under a blistering sun in old Cascade and managed to recover nearly one hundred coins and some gold and silver jewellery, I missed the Chinese treasure.

It's probably still there, as it has been since that fateful day in 1899, in a grass covered site two or three feet down in the remains of the old foundation of a hotel once known as the Grand Central.

THE LOST MINE OF PITT LAKE

In this article concerning the legendary lost mine of John Slumach which reputedly lies somewhere in a forbidding mountain valley high in the Coast Range north of Pitt Lake, Canada West Magazine has researched more than fifty accounts and articles in the Provincial Archives at Victoria, B.C. and carefully perused previously unpublished letters pertaining to this long sought gold mine. Innumerable misconceptions, exaggerations and embellishments were encountered in our analysis. After carefully considering the mass of evidence, it is the considered opinion of the writer that this highly colourful and interesting tale, accepted as fact by a host of treasure seekers, probably does not exist.

There is, however, enough intriguing and unexplained evidence that this story cannot be casually discarded.

It is this evidence which lends at least a modicum of plausibility to a fascinating tale which has drawn at least eleven prospectors and treasure hunters to their deaths in the past seventy years.

so I struck out whare
I wanted and I found the
Proof whare the creek
is by the formation

R.A. "Volcanic" Brown, May 28, 19[illegible]4.

In the first pages of this story the particulars of the legend of "The Lost Mine of Pitt Lake" will be presented. Although there are dozens of variations, the basic story has not changed for decades.

The story of "The Lost Mine of Pitt Lake" began in the late 1880's when one John Slumach, a Salish Indian from the Silver Creek Reserve, north-east of Vancouver, arrived in New Westminster, B.C. with a leather poke full of raw gold. According to traditional accounts, the gold was predominantly rough-edged, of good size with quartz adhering to some of the larger nuggets.

Slumach apparently attracted considerable attention in the Queen City with his rich burden. Evidently, however, amid growing speculation as to the origin of the gold, the Indian dissipated his assets in the saloons of the city and disappeared. Over the course of two years, the Indian purportedly paid two more visits to New Westminster and each time he brought a poke of nuggets. The value of each of the pokes cashed in was supposed to have been between $1,400 and $1,800. As before, Slumach rapidly squandered his funds, and despite several efforts to follow him, again vanished. It was at this time that rumours began to circulate about the old Indian's "Lost Mine."

Shortly after Slumach's last appearance in New Westminster, his name suddenly appeared in the news. The Indian was charged with murder!

The details of Slumach's crime, capture and eventual trial may still be obtained from the Archives in Victoria. Briefly, on September 8, 1890, Slumach shot and killed one Louis Bee, a half-breed who had insulted the

An early view of Columbia Street in New Westminster.

old Indian. The murder was committed at Lillooet Slough, about two and one-half miles above the Pitt River bridge.

After the crime, the woods wise Indian took to the mountains where he evaded capture for nearly seven weeks before finally surrendering to two Indian constables and P. McTiernan, the local Indian agent.

On November 3, 1890, a preliminary hearing was held in the city of New Westminster before Captain George Pittendrigh, J.P., where Slumach was remanded to the Fall Assizes.

On November 15, 1890, with Mr. Justice Montague W. Tyrwhitt Drake of Victoria presiding, the case was brought to trial. The evidence against the prisoner was damning and the jury, after deliberating only fifteen minutes, returned a verdict of guilty. Mr. Justice Drake condemned the old man to death by hanging and on January 16, 1891, John Slumach, man of mystery and condemned murderer was hanged on the gallows in the yard of the provincial jail at New Westminster.

A column in the Jan. 16, 1891 issue of the New Westminster COLUMBIAN stated:

> PAID THE PENALTY - Old Slumach was hanged in the yard of the provincial gaol this morning at 8 o'clock, for the murder, on September 8th last, of Louis Bee....To Mr. McTiernan, Indian agent, Slumach made a statement...in which he said the murder of Louis Bee was the only crime of the kind he ever committed, although he was credited with more....

At the time, numerous statements were attributed to Slumach. The Indians swore that Slumach stated tersely, "nika memloose mine memloose" - in essence, "when I die, mine dies," before he went to the gallows. There was, however, no official confirmation of such a statement.

Soon after the death of Slumach, reports of his lost mine began to circulate again, especially in New Westminster. These stories attracted a variety of individuals to the Queen City. One such man was W. Jackson, an experienced American prospector who appeared in the area in 1901. Jackson, a taciturn and methodical man, remained for several days in the city, listening to the various accounts and poring over the few available maps of the Pitt Lake region. Finally, apparently satisfied as to the authenticity of the tale, he purchased provisions and left for the Pitt River country, eventually vanishing into the trackless mountains north of Pitt Lake.

Several months elapsed before Jackson appeared again in the city. He was a changed man; haggard and emaciated and extremely close-mouthed. Pausing only briefly in New Westminster, the prospector headed south, taking with him a heavy pack which he guarded constantly. Subsequently it was reported that Jackson had cashed $10,000 in raw gold at a branch of the Bank of British North America in San Francisco.

The journey into the Pitt Lake country had, however, so adversely affected the prospector's health that he never fully recovered and during the remaining years of his life his illness worsened. In 1904, just prior to Jackson's death, he decided to divulge the location of the lost mine and wrote a letter to a certain Hill who had grubstaked him at one time. There are still some copies of this letter extant; the majority contain a vague description of the gold creek and the surrounding terrain. The copy published in this article was found in some 1924 correspondence of R.A. Brown, a noted prospector of Southern British Columbia for nearly fifty years.

I had been out over two months and found myself running short of grub. I had lived mostly on fresh meat for one cant carry much of a pack in those hills. I found a few very promising ledges and colors in the little creeks but nothing I cared to stay with. I had almost made up my mind to light out the next day. I climbed up to the top of a sharp ridge and looked down into the canyon or valley about one mile and a half long, and what struck me as singular, it appeared to have no outlet for the little creek that flowed at the bottom. Afterwards I found that the creek entered a ---------- and was lost. After some difficulty I found my way down to the creek. The water was almost white, the formation for the most part had been slate and granite, but there I found a kind of schist and slate formation. Now comes the interesting part. I had only a small prospecting pan but I found colors at once right on the surface, and such colors they were. I knew then that I had struck it right at last. In going up stream I came to a place where the bedrock was bare, and there, you could hardly believe me, the bedrock was yellow with gold. Some of the nuggets was as big as walnuts and there were many chunks carrying quartz. After sizing it up, I saw there was millions stowed around in the little cracks. On account of the weight I buried part of the gold at the foot of a large tent shaped rock facing the creek. You cant miss it. There is a mark cut out in it. Taking with me what I supposed to be ten thousand dollars (in gold) but afterwards it proved to be a little over eight thousand dollars. After three days hard travelling, it would not have been over two days good going, but the way was rough and I was not feeling well, I arrived at the Lake and while resting there was taken sick and have never since been able to return, and now I fear I never shall. I am alone in the world, no relatives, no one to look after me for anything. Of course I have never spoken of this find during all this time for fear of it being discovered. It has caused me many anxious hours, but the place is so well guarded by surrounding ridges and mountains that it should not be found for many years, unless someone knew of it being there. O, how I wish I could go with you to show you this wonderful place, for I cannot give you any exact directions, and it may take a year or more to find. Dont give up but keep at it and you will be repaid beyond your wildest dreams. I believe any further directions would only tend to confuse it, so I will only suggest further that you go alone or at least only take one or two trusty Indians to pack food and no one need know but that you are going on a hunting trip until you find the place and get everything for yourself. When you find it and I am sure you will, should you care to see me, advertise in the 'Frisco Examr.,' and if I am living I will either come to see you, or let you know where to find me, but once more I say to you, dont fail to look this great property up and dont give up until you find it.

Now good bye and may success attend you.

Yours truly,

W. Jackson

This letter has been reprinted verbatim from the original, complete with mistakes in punctuation and grammar. The copy was enclosed in one of the letters from Brown which was dated May 28, 1924. It is the only dated and authenticated copy in existence.

Hill failed in his attempts to locate the lost placer creek and eventually copies of Jackson's letter fell into other hands. There were reports of an accompanying map but there is no proof of the existence of such a map, certainly in Jackson's alleged letter there was no mention of an accessory map.

For the next quarter of a century an assortment of both seasoned prospectors and foolhardy novices tried their luck in locating the lost gold placer. Undoubtedly the most illustrious of the prospectors was one R.A. Brown, a grizzled and garrulous character from Grand Forks, B.C., who was generally known by the sobriquet "Volcanic," although the nick-names of "Sunset" and "Crazy" had also been bestowed upon him during his checkered prospecting career of over fifty years.

Brown began his search for the lost mine in 1923 after obtaining a copy of the Jackson letter and other pertinent information, all of which he considered irrefutable evidence. Each summer for the next seven years the resolute old man trekked into the rugged Pitt Lake country in his quest for Slumach's mine and each fall he returned, convinced that he was finally closing in on the fabulous gold stream. In September of 1930, however, Brown failed to return from the mountains. In November a search party consisting of Game Warden George Stevenson, Provincial Police Constable Eugene Murphy and two skilled trappers, Bill and Roy McMaster, set out. Due to injuries, Constable Murphy and Bill McMaster were forced to withdraw early. Stevenson and Roy McMaster continued the search for the missing prospector. For twenty-seven harrowing days the two pressed their search beyond the lake. Finally, near the headwaters of the Stave river, they came across Brown's last camp. The camp was abandoned and nearby lay a collapsed tent, a shotgun, a notebook with herb cures in it and some cooking utensils. Further inspection turned up an unusual discovery, a small glass jar containing eleven ounces of raw gold. Interestingly, the gold bore minute traces of quartz. Of Brown there was no sign.

The discovery of the gold in Brown's camp later gave rise to much speculation. Had he finally found the lost creek or the mother lode? The chance discovery of the gold was, unfortunately inconclusive, the gold may indeed have come from the lost mine area but it may also have come from another source outside the Pitt Lake region.

In the years which have elapsed since Brown's disappearance, others have tried to locate the lost placer. Some of these bids were mounted by individuals who were experienced, determined and well financed, others were solitary efforts by rash tenderfoots. All ended in failure and some in tragedy with the list of victims increasing with each passing decade.

Today there are many who maintain that Slumach's elusive Lost Placer Creek still lies deep in the inaccessible mountains north of Pitt Lake and eventually someone will find it. Others, equally adamant, are convinced that the story is simply a fabrication which has been perpetuated and added to over the course of the years since Slumach's hanging. They point out that the region immediately to the north of Pitt Lake is geologically unfavourable for both placer gold creeks and lode gold deposits. They also assert that there is no definite proof that Slumach ever found any gold at all prior to his death in 1891.

So the question remains unanswered. Does the legendary Lost Mine of John Slumach and Jackson actually lie undetected in a remote mountain valley in the precipitous country beyond Pitt Lake? One of the imponderables is the validity of the Jackson letter. If the letter were genuine then the mine does, in fact, exist. If, however, the letter were a forgery then the entire story is probably sheer fantasy but a deadly fantasy indeed, one which has led at least eleven men to their deaths in three-quarters of a century.

In the latter part of this article the most pertinent information is reviewed in order to determine the probability of the existence of the lost mine. The clues are covered point by point.

SLUMACH - The existence of Slumach is verified by both the court records and contemorary newspaper accounts.

SLUMACH'S GOLD - There is some doubt about the veracity of the generally accepted theory that the Indian cashed in large amounts of free gold in New Westminster prior to 1891. However, it is generally acknowledged that as early as the mid 1890s stories concerning "Slumach's Gold Mine" were circulating in New Westminster.

SLUMACH'S LAST WORDS - No authentication of the Indian's last words can be found in either the newspaper accounts of the period or in the court records.

SLUMACH'S HANGING - Definitely verified. Stories of other murders by Slumach must be considered untrue.

JACKSON - It is disputable whether or not this American prospector really existed. There is no absolute proof although there are numerous and usually vague claims that he did make an appearance in the city of New Westminster in 1901. If the letter is genuine then Jackson must also have existed.

JACKSON'S GOLD - Again no irrefutable evidence. Jackson, however, must have found the gold he claimed to have recovered if his account in the letter were true. And it seems to ring true.

JACKSON'S LETTER - Undoubtedly the key to the entire story. The letter or copy appears to be genuine. Jackson, reputedly an American, did spell certain words like "colors" and "o" in the accepted American version. Other terms like "chunks" and "light out" are terms which were used by prospectors of that era. Words like "schist" and "bedrock" indicate that the writer was familiar with mining terminology. Another point in the letter was significant: "Frisco Examr." refers to the San Francisco Examiner, which in 1904, was one of the leading newspapers in the city of San Francisco. A close perusal of the letter or copy lends credence to the theory that it is indeed genuine. After poring over hundreds of letters, diaries and documents from that era we reluctantly came to the conclusion that the letter is authentic. There are too many indicators to decide otherwise. "Volcanic" Brown's copy of the original was dated and several important words had been deleted but it was authenticated by a provincial government stamp. Until this copy of the original turned up there was no actual visible proof of the existence of the purported "lost mine" letter. Now, this copy, over half a century old, proves the old allegation that such a letter did exist. If it is a fake, which is highly improbable, it is the most clever spurious copy we have seen.

JACKSON'S MAP - There is little evidence to support the theory that such a map ever existed. The "Jackson letter" did not mention any map or even infer that one existed. It should be assumed that any "lost mine" maps are not bona fide.

THE BANK OF BRITISH NORTH AMERICA - This was a Canadian bank which was established in 1835 and continued in operation until 1918 when it was absorbed by the Bank of Montreal. Unfortunately, there are no records available to check out the statement that Jackson banked over $8,000 in gold at the San Francisco branch.

BROWN - "Volcanic" Brown was a well known prospector from the southern part of the province. He resided from most of his life in the Grand Forks area of the Boundary country. Brown met with varied success in his prospecting career which lasted nearly half a century. His failure to develop the highly touted Volcanic Claim on the Granby River near Grand Forks was overshadowed by his later triumph when he staked and sold the famous Copper Mountain property in the Similkameen country. An experienced and dedicated prospector, he devoted the last years of his life relentlessly searching for the "Lost Mine of Pitt Lake."

BROWN'S GOLD - Both Stevenson and Roy McMaster, who found the eleven ounces of gold in Brown's last camp attested to its existence. Both of these men, whose veracity is generally not questioned, have since died and the gold in question has disappeared. The gold, however, may have been from a lode deposit and not placer gold.

THE GEOLOGY OF THE AREA - The 1965 Geological Survey of Canada of Vancouver North, Coquitlam and the Pitt Lake Map Area states that the entire region is unfavourable for both placer and lode gold deposits. This fact, of course, was well known to most prospectors and some have suggested that the placer gold found by Jackson came from considerably farther north than the head of Pitt Lake. This is, however, doubtful as Jackson, according to his letter took only three days to get back to Pitt Lake from the lost creek.

PROBABILITY OF THE EXISTENCE OF THE LOST MINE - In some respects this most famous of lost mine stories is highly suspect; in others, fairly logical. The case for and against is about evenly divided. According to the geological information available it would appear unlikely that a large placer gold deposit would be found in the region. If Jackson's letter is to be believed, however, he actually located a creek in the area and recovered an unbelievable amount of gold from the bedrock of the stream, much of the gold heavily shot through with quartz. He also stated that up to the time that he came across the creek the region was not very encouraging geologically. The stream itself evidently flowed through a narrow mile and a half valley and disappeared underground at the end of the canyon. Jackson also stated that he had cached a large quantity of gold at the base of a huge tent-shaped rock which faced the creek. His explanation that he took out all the gold that he could carry also rings true. According to the price of placer gold at the time, he packed out nearly forty pounds troy, about all that a man in his weakened condition could have managed under the circumstances. It is an interesting story despite some contradictions.

The CANNONS of FORT RUPERT

An abandoned cannon still overlooks the bay near the site of historic Fort Rupert.

The gaunt stone chimney is all that remains of the fort. Once it had been a Hudson's Bay Company post known as Fort Rupert, a cornerstone of that company's far-flung fur trading empire on the north-eastern part of Vancouver Island. The dusty records in the provincial archives reveal its importance in those years when the company was expanding its activities in the west. There were much larger forts like Victoria, Langley, Kamloops, St. James and a score of others but few of them could match the defensive precautions taken at Fort Rupert. The original bastions bristled with cannon, the palisades, stoutly built, were high and heavily protected. The fort had been so cleverly constructed that covering fire could rake any point along the walls which were threatened.

These elaborate measures were essential because of the presence of the Nahwitti, a tribe of Indians who inhabited the surrounding area. The Nahwitti, proud and truculent, were reputed to be the most ferocious of all of the Indian tribes on the northern part of the island. It was a reputation which had been well earned for their relationship with the Hudson's Bay Company was one of mutual distrust and marked by sporadic and occasionally bloody hostilities by both sides.

The fort had been built in 1849 and remained under the banner of the Hudson's Bay Company until 1873 when the lack of profitability and the unusually difficult relations with the Nahwitti caused the Honourable Company to abandon the post. The Factor, Robert Hunt, immediately turned free trader and purchased the entire complex. For the next sixteen years the fort was the focal point of local trade although Hunt was hard pressed at times to show a profit from his operations.

Finally, on June 6, 1889 the fort disappeared in flames. The landmark which had stood for forty years had vanished.

Strangely, much of the atmosphere of the old post lingered on. The Indian village, Ku-Kultz, which had stood nearby remained where it was. The charred remains of the fort were soon overgrown and before long the only evidence of the old post was its huge stone chimney, everything else had vanished. The Nahwitti, however, like Indians everywhere of that era,

The Nahwitti Indians stare suspiciously at the camera while the officers of H. M. S. Scout pose for the photographer and posterity.

aware of the value of metal, salvaged many of the metal objects from the blackened ruins. Knives, nails, hammers and anything else even remotely useful was taken. And the Nahwitti, remembering with anger the havoc the iron and brass cannon had created when their village had been shelled, also took the big guns. Somehow, probably with improvised block and tackle, the Indians managed to cart even the eight pounders away from the remains of the fort. The four pounders were somewhat easier to remove and before long at least half a dozen cannon stood in Ku-Kultz. Although they were useless pieces of armament in the hands of the Indians they seemed to give their owners, and the village, a certain prestige. As the decades passed the cannon remained, almost forgotten, in the old village.

Occasionally infrequent visitors to the area would register surprise when they came across the cannon in the picturesque old Indian village. Their inevitable questions would be answered by the older Indians pointing to the overgrown remains of the fort and their explanation concerning the acquisition of the cannon and the history of the post and the village.

So it went until World War Two when an Air Force base was established several miles from Ku-Kultz. Some of the servicemen who were stationed in the area learned about the cannon and shortly after the war was over the cannon began disappearing.

By the early 1960s only three cannon remained at Ku-Kultz. The end of that decade found only two pieces left. The last remaining brass gun had been taken to Vancouver on a fish-boat by a member of the Hunt family.

Fort Rupert in its heyday. The defensive precautions taken at this post were almost unbelievable. It had the usual corner bastions but also had an extra bastion above the main gate. The entrance was also protected by a palisaded walkway which forced all trading Indians to enter the fort by this walkway which was covered by numerous cannon.

Indian houses built on the remains of Fort Rupert - circa 1896.

The family plot of the Hunt family just beyond the old fort site.

This cannon, the only authenticated H. B. Co. brass piece in Canada, was immediately purchased by a British Columbia collector who was concerned about the possibility of this museum piece falling into foreign hands.

Today the only cannon remaining at Ku-Kultz still overlooks the bay where British men-o-war once used to drop anchor. But the area itself is still being searched by cannon hunters who have been drawn by tales about other cannon in remote bays, shallow waters or in the underbrush close by the historic fort site. Several years ago a group of scuba divers stumbled across six cannon on the sandy beach of an isolated bay in the region. Surprisingly, the cannon were protruding out of the sand, clearly visible some distance away. These and similar finds have served to convince many that there are still cannon near the site of historic Fort Rupert and the Indian village once known as Ku-Kultz on the northern tip of Vancouver Island.

The brass cannon mentioned in the story. Like all of the cannon at Fort Rupert, it had originally been a ship's gun. The inscription "Amsterdam" and "1866" indicate the name of the original ship and the year.

This article was written in order to inform the readers of some of the history and atmosphere of that fascinating part of the province, the northern tip of Vancouver Island. We hope that any individuals who are attracted into the area will be interested solely in the history of the area as the cannon mentioned in this story must now be considered museum pieces and should be treated as such.

JOLLY JACK'S LOST PLACER

Old Jack Thornton pictured in front of his home cabin on Boundary creek. This photograph was taken sometime around the turn of the century. It was from this cabin that Thornton set out to disappear somewhere into the hills nearby to mine a secret placer creek. The location of his hidden mine was never revealed by the old man and it went to the grave with him touching off a search which has lasted from his death until today.

BY N.L. BARLEE

All photographs used in this article are from the Annual Report of the Minister of Mines and from the files of Canada West Magazine.

This is the story of an almost unknown placer creek, a stream which has yielded coarse gold and nuggets for over a century; more than a hundred years of production and yet, strangely, there are no records of the creek in the Mining Reports. It is also a tale of treasure; of a lost mine which was found decades ago by an old prospector known as "Jolly Jack" Thornton, a location which, at his death, was lost again and has been a lodestone for prospectors and adventurers ever since.

The city of Greenwood in 1900 with Boundary Creek in the foreground. Jack Thornton's old cabin still stands six miles south of the town.

The old prospector talked slowly and somewhat ponderously. His eyes were fixed on me but his thoughts were decades away.

"The gold was coarse and heavy," he mused, "with nuggets of up to an ounce. Most of it was bright, although much redder than placer gold usually is, the remainder was heavily oxidized, giving it a dark, coppery tone." He paused for a moment, thoughtfully rolled a cigarette, lit up and then continued. "You know, Thornton had a reputation for being talkative and in many ways he was, but when the subject of his secret placer was brought up, he always clammed up. Nobody, not even his own family, ever found out where it was - he never dropped a hint." He stopped and then went on. "I can't tell you much more, except that the location of his hidden mine must have been only a few hours ride from his old cabin beside Boundary creek, because he could make the round trip in a day and almost always did."

I left the old man to his solitary panning and on my way back home I mused over our conversation. The details were the same - they always were. Every one of the old-timers who had seen the gold from Jolly Jack Thornton's lost placer were in agreement about its appearance; very coarse and some of it oxidized. And almost to a man they agreed that the location had to be within easy riding distance from Jolly Jack's cabin on Boundary creek.

As usual, some of the information was precise. Most of it, however, was vague and based on hearsay. Mentally I went over the major placer gold areas in the Boundary country. The Rock creek region was easily the most celebrated gold area and that famous creek and its tributaries like Jolly, McKinney and Baker were all gold bearing but the lost placer couldn't have been in that region because it was a full two days round trip from Jolly Jack's home cabin. The Kettle river was out of the question because the gold on the river was too sparse and far too fine. That left two regions worth considering; Boundary creek and Fourth of July creek and its tiny tributaries.

The old original Thornton cabin as it appears today. Although it still stands close by Boundary creek the cabin is in such poor condition that it will undoubtedly fall victim to the elements in the near future.

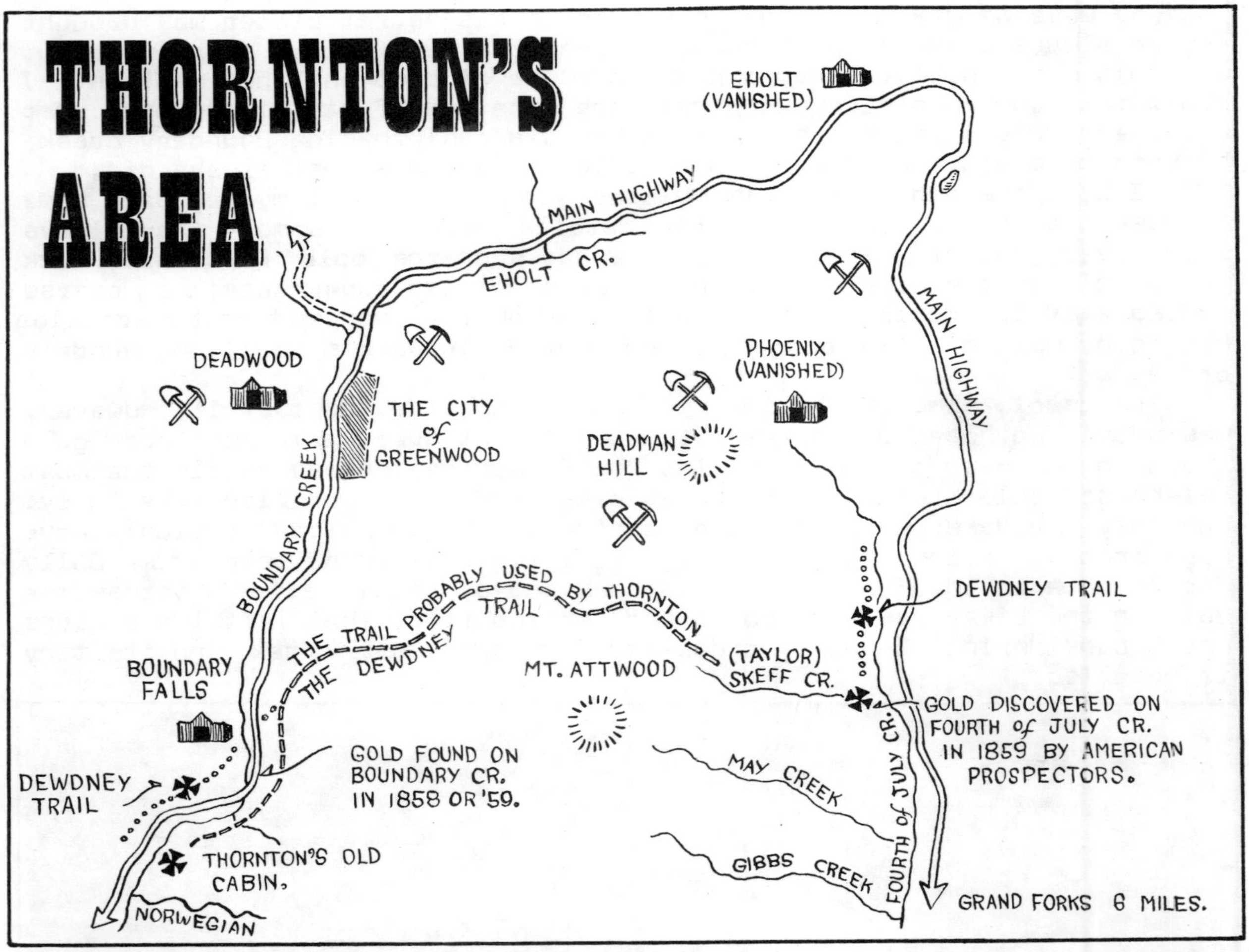

Jack Thornton had lived and mined on Boundary creek for years. South of the falls it had been exceptionally rich in places but that section of the creek was open and it would have been impossible for old Thornton, or anyone else, to have worked that stretch of the creek undetected, besides, the character of the gold from Boundary differed somewhat from the gold of the lost placer. That left only one area - Fourth of July creek and its few branch streams. Little was actually known about the history of that region except that placer gold had been discovered there in 1859 by a band of American prospectors. Some coarse gold had been found but the creeks had a reputation for being spotty and it seemed unlikely that any of them; Fourth of July, May or little known Taylor - any one of which a person can step across in low water, would be logical locations. I knew that a number of prospectors had scoured the Boundary country for years, searching for the elusive placer creek. Most of them had been experienced miners, and their sharp eyes seldom missed the tell-tale signs which indicated that a creek or area had been worked - but they too had come up empty-handed.

It was a perplexing and fascinating story, but one that seemed almost insoluble. I had half a notion that Jack Thornton's lost placer probably existed somewhere in the Boundary country but the few clues were less than encouraging, so, once again I dismissed the quest from my mind.

I had almost forgotten about it when I stopped by in Greenwood some years later to look up an old friend and one time prospecting partner of mine. His name was Peter den Hartog, a big man whose physical attributes had given rise to his nickname, "Big Pete." He was a thorough prospector who knew most of the gold creeks in the area intimately. We hadn't seen each other for years but I had heard in a round about way that Pete had come across a good placer prospect from which he was recovering a quantity of nuggets. Remembering the Jolly Jack story, I was mildly interested in the location of his find.

Later, while relaxing over a cup of coffee in his kitchen, I queried him about his discovery. Without replying he rose from the table and a minute later returned with a container. Carefully he spilled its contents out on the table. Half a pound of coarse gold and nuggets lay glittering in front of me.

"Bill," Pete stated matter-of-factly, "I think I might be on Jolly Jack's old spot."

I turned the gold over in my hand. It had a definite reddish hue, the exact colour that the gold from the lost placer was supposed to have, and much of it was oxidized. The characteristics certainly corresponded with the historical description of the lost creek gold.

"Where did you find it, Pete?" I asked.

"Skeff creek." he replied.

Skeff creek? I knew all of the gold creeks in the Boundary country but I couldn't recall any by the name of Skeff. I looked at the gold again.

"It looks almost like May creek gold." I ventured.

"Sure," he answered, "Skeff flows into Fourth of July just up from May."

Now I knew. Skeff was the new name for the little creek which used to be called Taylor. Many of the locals still called it Taylor creek.

I examined the gold again. The features of the metal matched all of the original accounts and the location of the area was logical, probably moreso than any other area in the entire Boundary country. Although the evidence was questionable there was still a possibility that Big Pete had actually come across old Thornton's long sought placer. By the time I left Greenwood several hours later I had decided that I had to check out that possibility.

For the next few weeks I pored over all the documented information about the lost mine that I could find. Some of it was of doubtful value, but some of it tended to support Pete's theory. Searching through my files I came across an interesting piece of information. In 1967, May Jones, one of Jack Thornton's surviving daughters had stated:

> 'My mother told me that my father really had found something very rich, but where it was he took to the Great Beyond with him. Before he got too ill he took his horse Rosie. It was in the spring of the year. He left home in the morning and was back at night and he had a baking powder can full of nuggets when he came home. He was old and so was his horse so he could not have gone far. She said he told her that they had struck it rich at last, but he never told mother where it was, or anyone else. Many have looked for it but none ever found the place yet.'

I perused the account again. His daughter had mentioned that the old man had made the trip in the spring of the year. It was a telling point because Skeff was, and still is, an insignificant creek which often dries up in the summer or early fall. The only time it flows with a volume of water adequate for sluicing is in the spring - it would be in the spring, and only then, that an experienced miner like Jack Thornton would work his secret placer. On that important point Skeff conformed.

Another key phrase stood out. Both Thornton and his horse Rosie were advanced in years when he worked his mine and indeed, as his daughter had stated, neither of them could have travelled far. I took out a detailed map of the Boundary country. I knew exactly where Jolly Jack's old cabin was on Boundary creek and I wondered how far it was from his home cabin to Skeff creek. There were several trails which the old prospector could have taken without being observed. The most logical route, however, was a seldom used trail which passed through the high country south of the old mining town of Phoenix, between Mt. Attwood and Deadman Hill. If he had taken that trail the distance between his cabin and Skeff creek was only seven and a half miles. Thornton knew the country extremely well and even at his age and on an old horse it would not have been difficult for him to leave home in the morning, work a rich pocket somewhere on Skeff creek, and be back home by evening.

So far the evidence was most intriguing, but there were still several

Jack Thornton undoubtedly travelled through this country just north of his home cabin to get to his unknown placer creek. The old horse trail between Mt. Attwood and Deadman Hill cuts through the distant hills visible on the extreme left of the photograph.

unanswered questions and only an intensive survey of Skeff creek could provide the answers. By early July I was on the stream. At first glance it didn't look impressive. The creek was exceptionally small. I had been on hundreds of gold creeks over the years but I couldn't recall any other which looked less promising.

I gradually worked upstream, looking for any indicators which might substantiate the lost placer theory. There were signs all along proving that miners had been at work in years past. The banks of the stream showed increasing evidence of placer operations as I moved upward. I passed by several abandoned sluice boxes, a broken rocker and the decaying remains of several ancient log cabins. Entering a narrow defile I noticed piles of tailings. 'Tailings' is a term used by placer miners to describe boulders which were removed to recover the gold lying on bedrock. The rocks had been painstakingly stacked on the banks of the creek so I knew that Chinese miners had once been there - white miners were invariably more slipshod than their Chinese counterparts, and they seldom piled their tailings as carefully. I was disappointed because I was aware that any creek that the Chinese had mined was usually worked out. Although they sometimes missed paystreaks, it was not a common occurrence.

Somewhat disheartened I continued on up the draw and passed under a sagging wooden bridge. Just beyond the bridge I paused - the old Chinese workings had suddenly stopped. For some inexplicable reason the oriental miners had worked up the creek to that point and then had simply quit. They had obviously abandoned the creek at that spot.

Why hadn't they continued? I knew from my research that there were only a few reasons why Chinese placer miners abandoned a creek. Either they ran out of gold, ran out of water or were driven off their ground by white miners. I knew that they hadn't run out of gold because Big Pete and others before him had obtained both coarse gold and nuggets just a few yards farther up the creek, so that wasn't the reason. They could have run out of water in the dry season, but when a stream was worth working, and Skeff certainly was, they would have done their dead work in the dry months and then mined the creek in the spring when the water was high. I crossed off that reason too. And historically, the Chinese had not been driven off any of the placer creeks anywhere in the Boundary country, so that wasn't the answer either. There had to be another explanation.

Perplexed, I gazed at the surrounding terrain. The stream continued to climb and on both banks massive boulders loomed up ominously. Some of them were thirty tons or more. Then the answer came to me - one of the Chinese miners must have been killed undercutting one of the boulders. Undercutting and propping boulders to get at the gold underneath them is one of the most dangerous operations in mining, and although the Chinese were experts at it, they too occasionally made mistakes. In those cases when one of their countrymen was killed, especially on a smaller stream, they always abandoned the locality, usually within hours of the mishap. That had to be it - it was the only logical explanation.

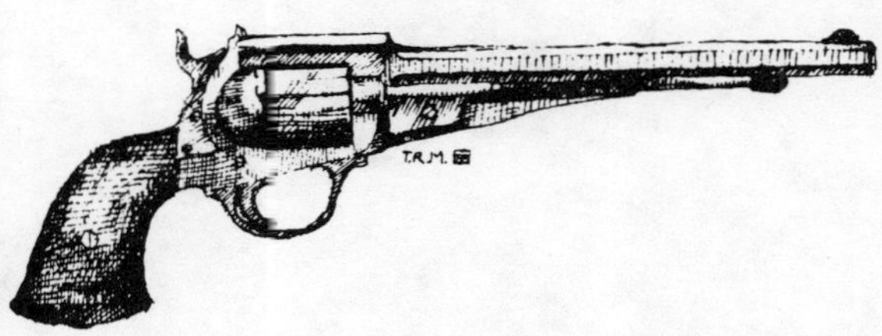

So the Chinese had come and gone, but others, lured by the coarse gold, had mined the creek after them. A few of them had done very well, especially those who stayed with the creek in spite of its difficulties. One, old Cal Hopper, had leased the ground and worked the creek for years. It was common knowledge that despite his frequent and expensive drinking bouts, he always had three or four medicine bottles full of coarse gold on hand in his cabin. Hopper died in 1943 and his lease passed into a succession of hands, none of whom was aware of its intriguing history. The coarse gold continued to turn up in tantalizing quantities but the boulders and overburden which had hampered former operations continued to plague them, and after tapping the creek for some of its riches, they gave up or moved on to easier ground elsewhere.

Surprisingly, the creek lay nearly deserted for almost two decades, only a few panners or disinterested miners trying their luck occasionally. When, in 1972, the lease on Skeff was finally abandoned, Big Pete acquired it. For the next two summers he systematically mined a small section of the ground and recovered ounce after ounce of gold. Pete worked the claim only in his spare time but the longer he was on the creek the more convinced he became that he was onto something out of the ordinary. By comparison with any of the other placer creeks he had mined, the gold from Skeff was not only more plentiful but also much coarser - almost half of the gold found being of nugget size.

For the remainder of the summer I made numerous trips to the claim, familiarizing myself with the ground and the history of the immediate locality.

I discovered that the surrounding area, although seemingly ordinary, was steeped in lore. The stream rose in the mountains near the historic

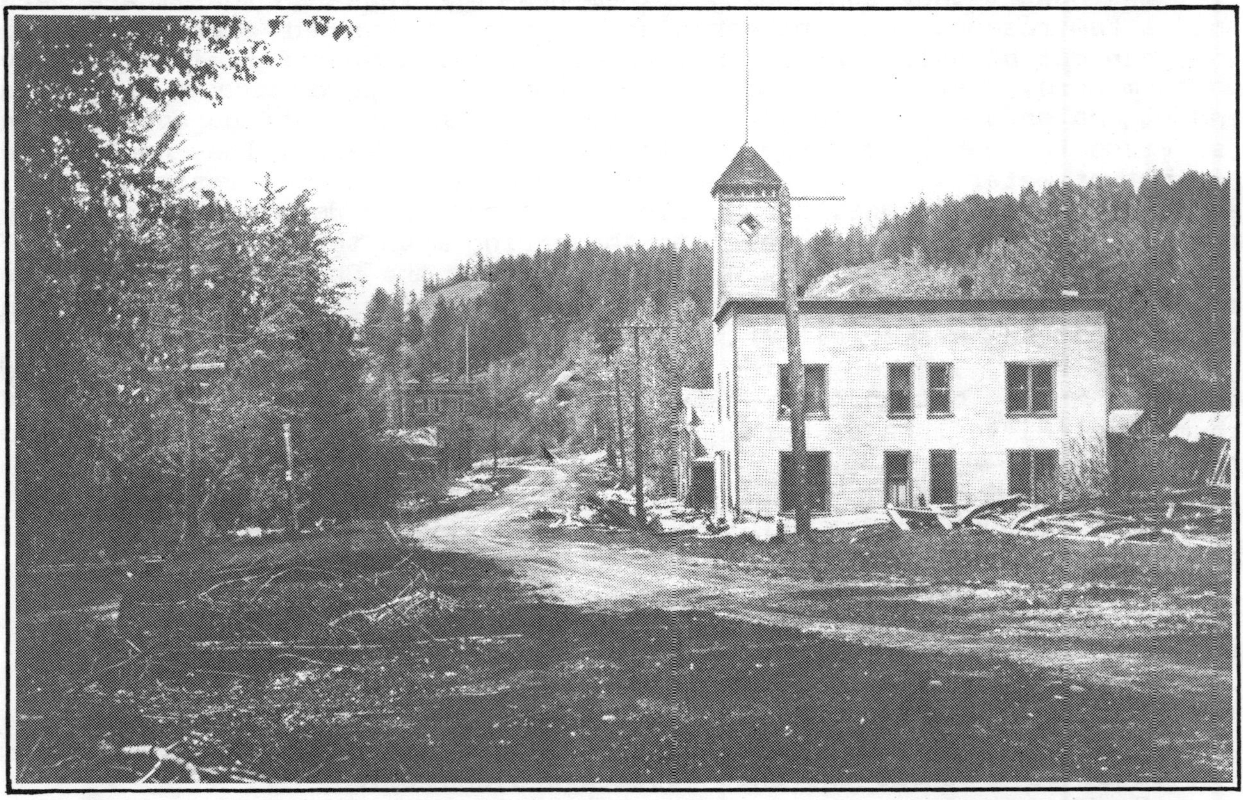

The Phoenix city hall in the late 1930s. When Thornton passed this town at the turn of the century it was a flourishing city. It died in 1919.

old city of Phoenix; a mining camp which had become an incorporated city in 1899, a ghost town twenty years later and finally had vanished from sight. And the famous Dewdney Trail, completed in the 1860s to link Fort Hope on the Fraser river with the renowned placer diggings at far off Wild Horse creek in the East Kootenays, followed Fourth of July creek and cut across the mouth of Skeff only half a mile from the main workings. Close by the trail were other reminders of the past; sod roofed log cabins, long abandoned and forgotten over the passing years, and evidences of old tragedies like the overgrown graves of three miners, victims of a vague disease known as "trail fever," who were buried close by that route of a century ago. I gradually realized, as time passed, that there were events which had taken placer in the area which I would never be able, in spite of all my research, to reconstruct.

So, as summer came to a close and the sharpness of autumn set in, I weighed the evidence I had accumulated.

Historically, the gold from Jack Thornton's lost placer was unusually coarse, copper tinted and a significant percentage of it was highly oxidized. The gold recovered from Skeff creek matched those characteristics exactly. It was a crucial point. I contemplated the other salient factors. All of the original records and available accounts had stated that Jolly Jack had made the trip from his home cabin on Boundary creek to his hidden mine in a single day; leaving in the morning and returning as night fell. Skeff creek was only seven and a half miles from the old man's Boundary creek cabin, a distance which would have allowed ample time for the old prospector to ride out in the morning, work a rich lead on Skeff and be back home by evening. Skeff was the only stream close enough and which carried gold coarse enough to fit those specifications. One nagging thought, however, still bothered me. Why hadn't other miners worked

Skeff creek after the Chinese vacated it? But there seemed to be a logical explanation for that too. Other miners probably assumed that the Chinese had left because the creek had been worked out by them, whereas Thornton, being an experienced hand, had decided to investigate and probably found hard evidence that the orientals had abandoned the stream because of a fatality. It stood to reason that he would then have thoroughly prospected farther up the creek, above the Chinese workings, and in so doing, may have come across a rich pocket which he subsequently mined. He would not and could not have worked the lead in the summer because of traditionally low water and thus visited his mine only in the spring as all reports had stated. At that time he could have sluiced his ground and then camouflaged his mine until he returned the next spring. The reason the creek hadn't been thoroughly worked out after Thornton's death was self explanatory, later miners simply weren't aware of the history of the ground and never arrived at the conclusions that the cagey Thornton had. The reasons why Jolly Jack hadn't disclosed the location of his hidden placer mine to his wife or other members of his family were also logical. Like nearly all miners, even when they are ill, they are positive that they will recover despite the symptoms. Thornton was old and ill, but he was tough too, and undoubtedly assumed that he would survive to work his hidden mine once more. Fate, however, intervened and the secret went to the grave with him.

After reviewing the evidence I finally came to the conclusion that Jolly Jack's long sought placer mine had undoubtedly been somewhere on Skeff creek, probably above the original Chinese ground. The evidence was just too overwhelming to ignore. Everything fitted; the distance to Skeff creek from the Thornton cabin, the coarseness of the gold, the tint, even the lack of data which had plagued researchers for years. There wasn't a single mention of either Skeff or Taylor creek in any of the old British Columbia Mining Reports from 1874 to 1973. Nobody, outside of a few local people, even knew that the stream had ever been a placer producer.

There is still gold on Skeff creek - coppery hued and very coarse - it's more difficult to recover today, the overburden is deeper and the boulders still guard their hoard as they have for more than a century. But, when the twenty, thirty and fifty dollar nuggets turn up and the gold lies glinting on the bedrock, the old mine always seems close at hand. And there is almost half a mile of ground along the creek which has never been touched by any miner. It's there, just up from the historic Dewdney Trail and south of that landmark known as Deadman Hill. Somewhere along that stretch, above the old Chinese diggings, Jolly Jack's lost mine may still lie hidden, awaiting rediscovery.

Some of the nuggets recovered from Skeff creek. They range in value from $5 pieces to $70 slugs. The darker nuggets are oxidized. The gold shown is unusual because of its coppery-red colour.

The FRONT STREET COINS of OLD YALE

The sternwheeler R. P. Rithet at Steamboat Landing in Yale in the 1880s.

Front Street is practically deserted today, grass grows where the boardwalks once were, and here and there, a stone wall or an overgrown depression marks the location of a business which flourished years ago.

Below the street, on the edge of the Fraser River, Steamboat Landing lies silent as the waters of the river flow endlessly by. Ring-bolts are still anchored to the huge boulders as they were so many years ago when the sternwheelers called regularly at Yale - the "Head of Navigation."

The whistles of those graceful vessels are no longer heard as they round the last bend on their final lap to the town. But over a century ago Yale was "the" town in the interior of British Columbia. Its main street, Front Street, was wide open seven days a week, twenty-four hours every day and it was teeming with people, mostly free spending miners and later, equally boisterous railroaders. According to contemporary accounts it was, in those years, one of the busiest thoroughfares in the entire west. And the catalyst of Yale's good fortune was that noble metal, the element called gold.

In 1857 it was common knowledge locally that placer gold had been found in quantity in "Couteau Country," that is that vague area east of where Lytton now stands. Late in the year the news of the discovery drifted to California. With the placer ground of that state on the wane, it didn't take much to convince the miners of the Golden State to head north. By the spring of 1858 they were arriving in British Columbia by the thousand. Some of them had come overland by various routes, others, the majority, by ship to Victoria. Soon a vast body of miners ascended the Fraser River, panning the bars as they moved upriver. At first the bars of the Fraser were disappointing, but as they continued on the more experienced miners noticed that the gold was increasing in both quantity and size. It was an encouraging sign, so they pressed on. Just below Fort Hope they encountered good diggings at Cornish and Maria bars and past that frontier post they began to hit much richer ground as bars like Murderer's, Emory, American, Canada, Santa Clara, Texas and a dozen others yielded astonishing amounts

The church of St. John the Divine overlooks scenes of debauchery on old Front Street in Yale during the railroad construction days. The street scene by an unknown artist is probably authentic although the names of the businesses pictured seem to be inaccurate. The old church is still standing today but Front Street, that rollicking way of yesterday, is almost deserted. The buildings pictured in this sketch were levelled by a series of fires.

Historic Hill's Bar today.

Leg irons and a spur from Hill's Bar.

of the yellow metal.

The main body of miners was steadily working upriver towards Yale and the mouth of the Fraser Canyon. Yale, which had been built only ten years earlier by the Hudson's Bay Company, had been ill-starred from its beginning. Only a year after it had been erected it had been abandoned by the Honourable Company. Now, with the influx of miners streaming upriver, it awaited them and a second chance.

By sheer good fortune, only a mile and a half below the fort, on the opposite side of the river lay a prominent gravel bar marked by a huge boulder protruding from the gravel at the river's edge. The vanguard of prospectors stopped and tested the gravels. Amongst that group was an old hand named Hill. It was he who panned out the first coarse "colours" on the bar which was promptly dubbed "Hill's Bar" in his honour. That bar, from that discovery day in 1858, to today, holds the record as the richest gravel bar in the history of the Canadian west. Half a mile long, sixty feet wide and five feet deep, it yielded staggering amounts of gold. It was so rich that some miners, who had been able to claim good ground on the bar, never worked again after they had mined out their claims. It has been estimated that over three tons of placer gold came off this bar in its life.

The discovery was not only a bonanza for the miners but also for the fort. Almost overnight it mushroomed into a bona fide boom town. False-front buildings were thrown up as soon as the lumber was whipsawn. Most of the businesses were erected on a high bench overlooking the river. It was only natural that this street was called "Front Street." That summer

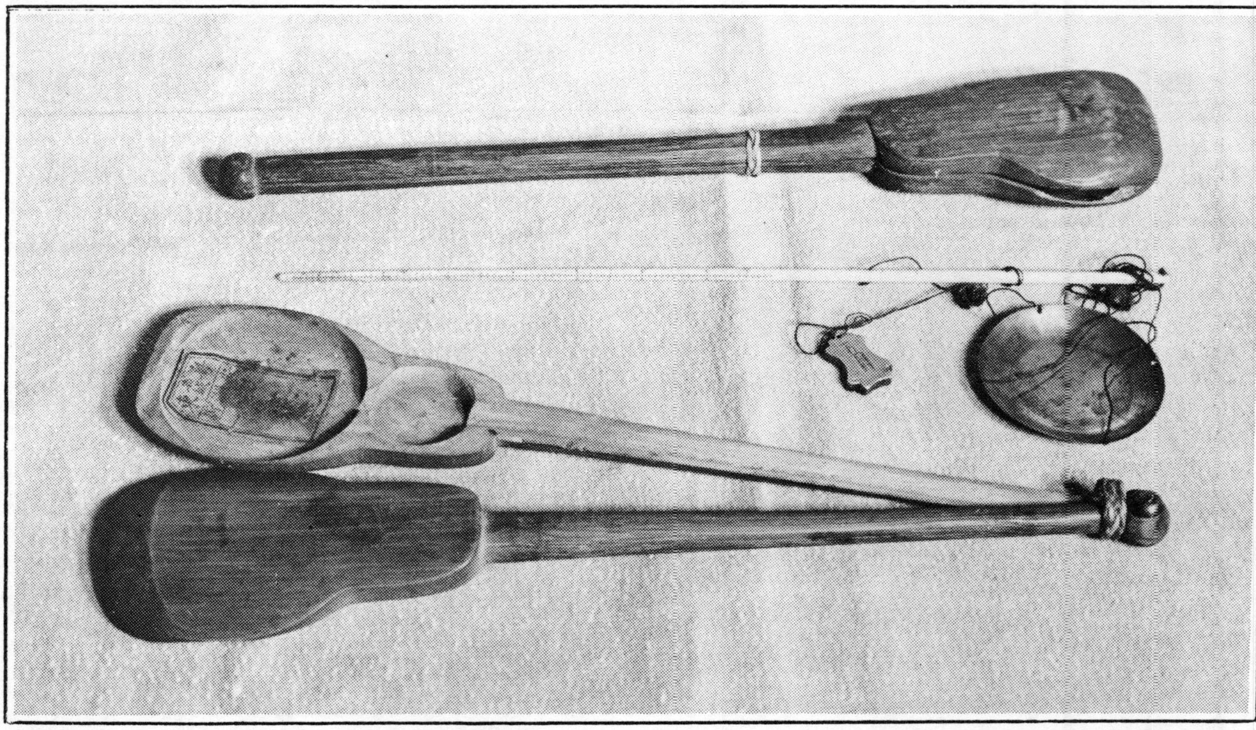

Two sets of Chinese gold scales from Yale. These sets, made of ivory and brass with bamboo cases, are more accurate than their western equivalents.

Yale really came into its own when the first sternwheeler, an American vessel called the "Umatilla," reached the town. The arrival of the ship was greeted with wild enthusiasm; pistols and rifles were discharged by excited miners lining the banks of the river. When the steamer rounded the bend below Yale and hove into view, rejoicing prospectors waded out to meet her.

Other vessels soon followed and Yale's fortunes continued to climb. By the end of that year it had a population of nearly 3,000 with another 1,000 miners on the bars nearby. The Fraser, near Yale, was referred to as "Ounce Diggings," and as such attracted all kinds of individuals. Like many towns on the edge of the frontier it attracted more than its share of misfits and adventurers. Some of them, like the notorious Ned McGowan, J. W. Bagley and Terrence Kelly, had vacated the Golden State barely a step ahead of the hangman's noose. The Vigilance Committee had publicly threatened to "execute them without trial" should they ever set foot in certain parts of California again.

By the end of 1858, more than 20,000 men had passed through the town on their way upriver. Because of this seemingly endless procession, the coffers of the businesses of Yale were filled to overflowing.

By geological necessity Yale was the gateway to the upper river and by the early 1860s it was firmly established as the only major community of any consequence above New Westminster although a gold town far to the north, called Barkerville, was soon to come to fame. In 1861 the Cariboo Road was started at Yale and by the fall of 1863 it extended 259 twisting

Part of a cache of over 300 Chinese coins found in Yale.

miles to Soda Creek. It was a boon to the town for all through the golden era of the Cariboo the traffic continued to funnel through Yale. On busy Front Street the crowds jostled shoulder to shoulder while freight wagons, stagecoaches and an assortment of other conveyances arrived from or left for the distant Cariboo. At Steamboat Landing the sternwheelers, sometimes several at once, deposited their cargoes and then headed back downriver to replenish them once more.

Those first years were heady days. The town was doing such a roaring business that there were times when people thought it would never end. Its prosperity, however, relied heavily on the Cariboo and when the gold returns from that region began to slip, Yale also felt the pinch. By the seventies the town was in the doldrums, the population had dwindled to under a thousand, businesses were being boarded up with alarming regularity and the once frequent sternwheeler arrivals were drastically curtailed. To add to its misfortunes several devastating fires had swept along Front Street, levelling every building in its path. After each holocaust Front Street had been rebuilt. Horse drawn scrapers would scrape away the debris and dump it unceremoniously over the bank. In several of the fires almost everything had been lost in some businesses: coins, jewellery and other valuables. These items, along with everything else in the blackened ruins were simply deposited over the edge of Front Street. This performance was repeated each time a fire razed the street.

By 1875 it appeared certain that Yale would suffer the fate of most towns founded on the shifting fortunes of gold. In 1879, however, Andrew

Barnard's Express office on Front Street in Yale in the 1860s.

The remains of Front Street after a fire of the 1880s.

Onderdonk was awarded his first contract by the C. P. R. and suddenly Yale roared back to life as hundreds of railroad workers invaded the town. When more contracts were awarded to Onderdonk in the early eighties, the tempo of construction continued unabated as the town remained the main supply point.

With the upsurge in Yale's fortunes, the riverboats returned in force to a chorus of salutes from the anvil batteries. The Pioneer Line, with sternwheelers like the William Irving, R. P. Rithet, and the ill-starred Elizabeth J. Irving soon monopolized the river trade. Along Front Street and in upper town plush establishments like the Railroad House, Caledonian, Steamboat Exchange, York, California, Branch, Columbia House and numerous other hotels catered to the free and easy railroad men. The saloons, offering an enticing variety of beverages, reaped exhorbitant profits. Around the clock poker games were the order of the day on Front Street. In Chinatown, on the northern end of the street, fan-tan and other favourite games of chance lured the taciturn Chinese.

For six years Onderdonk's crews, both Chinese and white, blasted and built over some of the toughest terrain anywhere in the west. And Yale rode high until the rail links through the canyon section were completed in 1885. When the line was finished so was Yale. The railroad, which had been responsible for the town's reprieve, had

A few coins from the Jim Shilson Collection in Yale.

had sealed its fate. Within a year the town had lapsed into an economic backwater from which it never emerged. Although a few brief flurries of activity provided temporary respites, by 1920, even the town's stoutest supporters were forced to admit that Yale would never again approach its prominence of old. And it never did.

Today the atmosphere of the gold years somehow lingers in old Yale, especially along almost vacant Front Street. And it is common knowledge amongst a few old-timers in the town that the practice, in the early days, of scraping the Front Street stretch after a fire and dumping the residue over the edge of the bank resulted in an extremely valuable dump.

Evidently the wild spending miners of the late 1850s and the equally open-handed railroaders of the 1880s lost an astounding number of coins in their day, and more than a few were gold. The recovery of gold pieces from this section of the town is impressive. In the past few years the following gold pieces have been found: a U. S. $5 gold piece, an English gold sovereign, a U. S. $2½ gold coin, a rare California hexagonal 50¢ gold piece, another U. S. $5 gold coin. And the list is far from complete because many finds were probably never reported. The total known number of coins recovered runs into the hundreds, almost all of them from Front Street.

Hopefully, any future finds along this historic street of yesterday will be made by historians, or even archaeologists, groups who have too long neglected our gold rush history. After all, Yale's Front Street was once the main street of one of the most historic and colourful towns in the old west.

The LOST MORGAN MINE

Greenwood City in 1898. Harry Morgan passed through this old town on his way up Boundary Creek which is visible in the foreground. The long lost silver mine is located near the headwaters of this stream. (Photograph - Courtesy of the Provincial Archives)

"The Lost Morgan Mine" was originally published in the Fall 1972 issue of Canada West Magazine. It was the first time that this story had been published anywhere although it had been a poorly kept secret in old Greenwood for decades. The background research indicated that this reputed lost silver mine may actually exist.

His name was Harry Morgan and he was an American who had drifted in from Chesaw, a little stopping-place on the U. S. side of the line just across from Bridesville in the Boundary Country.

It was near the turn of the century when he hit Greenwood. In those years "The Boundless Boundary" was in the midst of a spectacular mining boom and all sorts of characters and promoters were in the area hoping to participate in the wealth. It looked as though every available piece of ground, no matter how remote, had been investigated by prospectors and over 800 claims had been recorded: all the way from Camp McKinney in the west to Burnt Basin in the east.

Morgan sized up the situation and quickly concluded that all the good ground had already been taken up, but as he was a trapper as well as a prospector he decided that he would eke out a living by trapping until he could latch onto a good mineral claim. With this plan in mind he left Greenwood and headed north. Following Boundary Creek he passed by silver rich Providence Camp, and a few miles farther on, Long Lake Camp, but he continued on until he neared the headwaters of the creek. Finally, about eighteen and a half miles north of Greenwood, he stopped. It was a remote region and it looked, to his experienced eye, like good trapping country. Within a few days he had erected a crude sod-roofed hut and had set out his traps.

Like so many other trappers and prospectors in the area he was seldom seen in town except when he was in purchasing supplies, and even then he attracted little attention because of his diminutive size and his unusual taciturnity: he was generally regarded as a strange man who had no known close friends. Several years passed with his usual twice-a-year trips as his only contact with Greenwood. One fall, however, he deviated from his

The city of Phoenix no longer exists, but it did in Morgan's time.

general pattern and turned up in Greenwood again just several days after he had left town. This time, instead of heading for a general merchant's store, he went to an assayer, a certain Captain Gordon. Gordon, one of numerous assayers in the district had a reputation for being very close mouthed. This was probably the reason Morgan chose him in preference to several others. Upon arriving at the assayer's place, Morgan untied two sacks which had been strapped to a pack-saddle of his black pony, and with some difficulty succeeded in packing them into the assayer's office. Once inside he tersely asked Gordon to assay the ore. Gordon, never having laid eyes on Morgan before, took the sacks from the prospector and spilled the contents out onto the table. He hefted a piece of the ore and noted that it was very heavy, his shrewd eye also perceived that the samples were chunky and had obviously come from a vein of considerable width, not a narrow deposit.

Nonchalently Captain Gordon agreed to run as assay on it to see what values, if any, it contained. Morgan looked warily at him and mumbled that he would return in a few days to find out the results. A moment later he was gone.

Gordon ran the assay that very night and was dumbfounded to learn that the ore carried between 400 and 500 ounces in silver and from five to six ounces of gold to each ton - it was staggeringly rich.

A few days later the prospector returned and listened impassively as the excited assayer read out the assay results. After he had finished the assayer mentioned that if Morgan hadn't staked the ground he had better do so in a big hurry. The prospector shook his head negatively and stated that nobody else would ever find his mine anyway.

According to local reports Morgan returned to Greenwood every few months with more samples for assaying and the results were always the same; extremely high values in silver and gold.

Finally, after several years association with the old prospector, the assayer, Captain Gordon, gained Morgan's confidence and was eventually invited to go out and take a look at the mine. He readily consented and a date and time were set. On that day Gordon left Greenwood and proceeded up Boundary Creek where he was to meet with Morgan at a pre-designated spot several miles south of the headwaters region. After travelling most of the day he arrived at the place agreed upon and found the old man waiting there. They set off together towards the headwaters but, as the assayer later related, the prospector seemed to display signs of nervousness the farther up the creek they travelled until finally, in Gordon's words "He just took off through the jack pines." And from that day on Captain Gordon never saw Morgan again, he simply vanished. The disappointed assayer made his way back to Greenwood and waited vainly for several weeks for the old prospector to turn up. Finally, possessed with curiosity he travelled into the area again to search for the old prospector. He found his sod-roofed cabin but no sign of the prospector or his mine which had produced the fabulous silver ore. Keenly disappointed he returned to town and related the details to several close friends. One look at the ore convinced most doubters and in the following months a number of other prospectors made efforts to locate "Morgan's Mine," but to no avail. Morgan had disappeared and his mine could not be found.

Some individuals made concentrated efforts to locate it. One old hand from Deadwood, Scott McRae, established a trap line in the area and spent years searching for it but drew a blank. Another trapper, Vic Barrett, set up his trap line in the same region after McRae had gone. Barrett was the

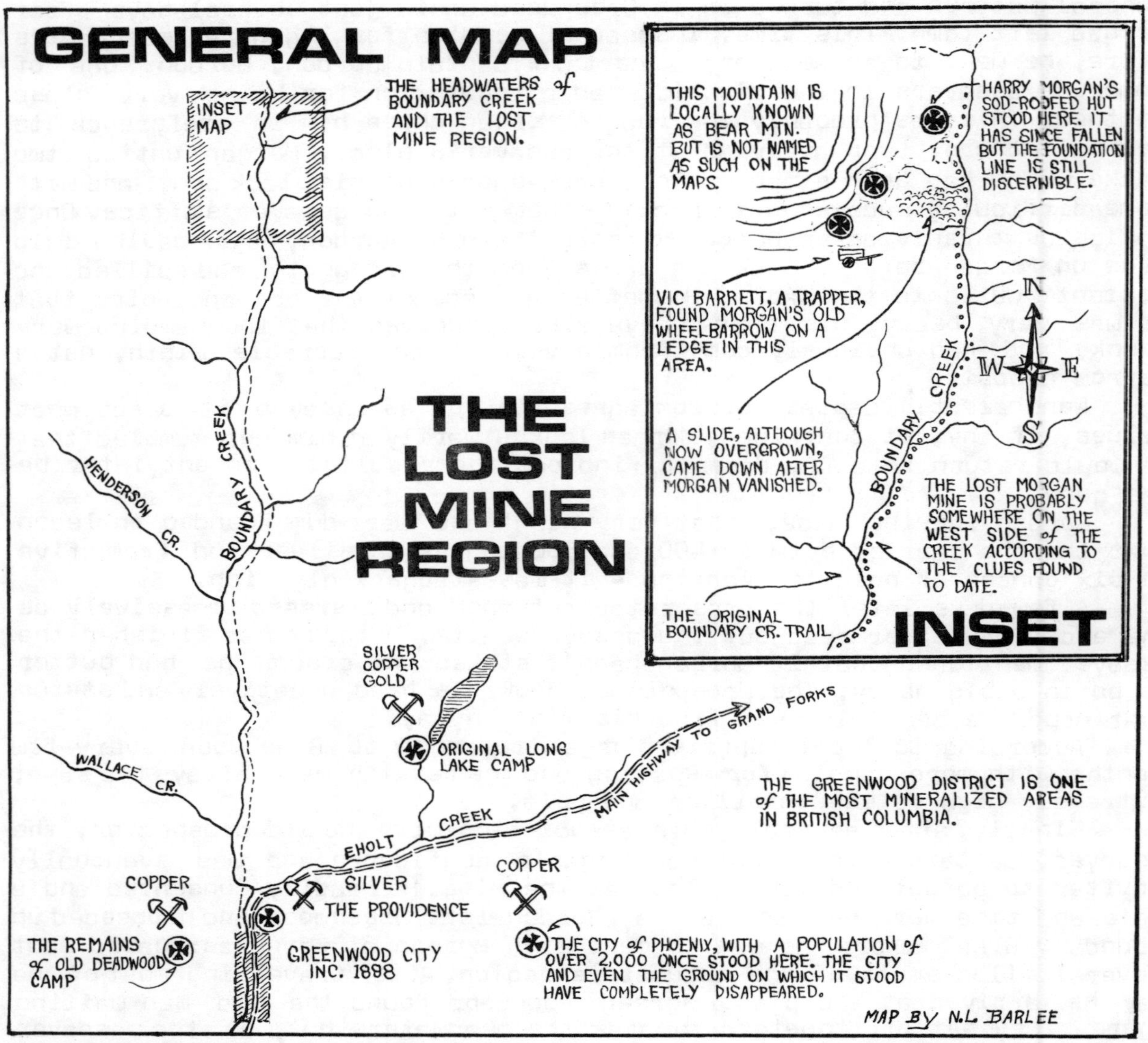

complete woodsman, strong as a horse and well versed in every aspect of woods lore. Barrett, however, was more fortunate than his predecessor for he finally succeeded in finding one important clue - Morgan's wheelbarrow, a crude affair which the old prospector had used to truck the silver ore out of his mine. The wheelbarrow had been found on a ridge about one and a half miles west of the creek and nearly three and a half miles from the headwaters. Sensing that he was finally hot on the trail, Barrett scoured the immediate area but was unable to find any sign of the mine. He finally gave up after days of searching and concluded that the wily Morgan had carefully camouflaged the entrance to the mine to prevent anyone else from finding it.

In later years Barrett led numerous expeditions into the area in search of the Morgan Mine but they were always unsuccessful. He continued into his late seventies and even up until his death in Greenwood on April 23, 1962, at eighty-six years of age, he was convinced that it was still near the headwaters of Boundary Creek awaiting rediscovery.

There were others who were equally certain, one was Fritz Hauzner who had a ranch up Boundary Creek. Hauzner had evidently procured some of the ore from Gordon, the assayer. It was phenomenally rich and bore out the first assays of the ore.

The formation near the headwaters is primarily granite-diorite but the Boundary Country, although noted for its copper producers, has had many celebrated silver mines. Some, like the Providence, Elkhorn, Skylark and Last Chance produced millions of dollars in silver ore in their day. There is good reason to believe that the Harry Morgan Mine is indeed in the immediate vicinity of the headwaters of Boundary Creek.

Where Morgan went or what happened to him is highly debateable. He may have been killed in a cave-in at his mine or he may have pack-trained some ore across the line and then decided not to return. At any rate the disappearance of Morgan has never been solved. The old records have been checked and they indicate that Morgan never staked the property but this was not unusual as many old-timers worked ground which they never staked formally.

The known facts are interesting. Harry Morgan did exist and he did have a trapline and cabin near the headwaters of Boundary Creek. He also brought in amazingly rich ore samples to Captain Gordon, the assayer in Greenwood. The samples were astoundingly rich and assayed more than $500 to the ton in silver and gold values. The samples were large and chunky, indicating that they had been mined from a wide vein, a vein which probably contained a considerable deposit of high grade ore. Vic Barrett did locate Morgan's wheelbarrow which had evidently been used to truck out the ore, a number of locals saw the wheelbarrow on the ridge where it had been found by Barrett.

That the local inhabitants are convinced as to the mine's existence is borne out by an amusing incident which occurred several years ago.

It was a typically hot summer's day when a local timber cruiser walked into the only beer parlour in Greenwood to slake his thirst. Noticing that the premises were jammed with patrons, and realizing that nobody was about to budge from their places, he struck upon a novel idea. When the lull of conversation died momentarily he remarked to an acquaintance that he had just been up around the headwaters of Boundary Creek and had discovered the remains of an old wheelbarrow. At this point the silence in the room was really quite remarkable. He went on to say that he had also noticed an old caved-in shaft nearby, and being relatively new to Greenwood, he wondered if anybody had ever had a mine up in that area. Several moments passed, then, one by one, and with a truly astonishing variety of reasons and excuses, the patrons, as nonchalently as they could manage it, filed out of the beer parlour. The timber cruiser then sat down in a virtually deserted beverage room and ordered his ale. A local wag later remarked that it was positively uncanny how many people had decided to take a hike up Boundary Creek that day "to see the country up there."

Even today the old residents of Greenwood are reluctant to discuss the mine and it took a considerable amount of cross-checking and research to compile the data in this story. Most lost mine stories are not quite plausible but "Harry Morgan's Lost Mine" probably does exist, somewhere in that lonely, desolate country near the headwaters of Boundary Creek.

Canada West Magazine is indebted to the Corporation of the City of Greenwood and to various individuals who were kind enough to supply certain information concerning the old Morgan Mine.

GOLD BULLION AT TWELVE FATHOMS

Articles such as those pictured above were lost when the PACIFIC went down in 1875. Gold and silver specie, nuggets and miscellaneous other items still await the treasure hunter who can locate the hulk of the old sidewheeler.

BY N.L. BARLEE

There are innumerable identified and unidentified wrecks along the Pacific coast of North America. Some of these sunken vessels contain treasure and this article is about one of these lost ships.

Almost $700,000 in sunken treasure - and it's still lying there; the bulk of it in raw gold from the once-booming creeks of the distant Cassiar, and the remainder from the Cariboo.

In 1875, Victoria, the one-time Hudson's Bay Company fort, reigned as the undisputed metropolis of British Columbia; the former frontier outpost had turned into a cosmopolitan city, rivalled on the Pacific coast only by old San Francisco.

The gold rush to the Fraser River in 1858, with its sudden influx of thousands of miners, had provided the initial impetus and prosperity and when the Cariboo came into prominence, again Victoria prospered, for it was from that town that supplies and equipment came and it was to Victoria where the wealthier and more fortunate miners gravitated in the off-season to spend the winter in more hospitable surroundings, for a winter on the Cariboo creeks was less than inviting and few stayed unless they had no alternative.

As the 1860s drew to a close, however, it became apparent that the peak years of the Cariboo had finally ended and that that once formidable goldfield was on the decline. But in 1873 another great placer gold area was discovered as the "Cassiar" in northern British Columbia came to the fore. By 1875, Thibert, Dease and McDame, that renowned trinity of Cassiar creeks was pouring out its wealth in awesome amounts and nearly 2,000 placer miners were combing the gravels for their riches. By early fall of that year many of the argonauts had made their stakes, and with bulging pokes they left the north country and headed for Victoria.

The sidewheeler PACIFIC. This painting was probably commissioned in her earlier days before her tragic fate off Cape Flattery. Her bones still lie off the cape where she went down 97 years ago. And the treasure she carried, now estimated at nearly $700,000, is still there only 12 or 13 fathoms deep.

The routine seldom varied, a few, often boisterous days were spent in riotous living in Victoria, and then, if their purses had not been relieved of too much gold, they continued their interrupted journey. Many of the miners had originated in California and that was their destination.

Almost all of the prospectors booked passage on one or other of the numerous vessels leaving Victoria for San Francisco, and being miners they were not very fussy about which ship or even what kind of ship they left on. In the fall of that year, one of the ships lying at anchor in Victoria Harbour was the PACIFIC, a vessel which had plied the turbulent waters of the Fraser River during the gold rush of 1858 but had been retired from active service in 1872 due to her general unseaworthiness. In 1874, when the excitement over the Cassiar was at its height she was pressed back into service, as was nearly every other ship able to float.

On November 4, 1875, she began taking on passengers for a scheduled run to San Francisco. It was a busy time of the year because of the many argonauts heading for the Golden State. Tickets were at a premium and many impatient miners who were lacking tickets simply boarded the creaking craft at the dock, thus dangerously overloading an already crammed ship. Besides the usual crew complement of nearly fifty, there were, according to eyewitnesses at the time, more than two hundred passengers on board the unstable vessel. The precise number, however, will never be known due to the absence of any complete records.

Finally, with her decks black with passengers, the ill-fated vessel steamed out of Victoria Harbour - on what was to be her last voyage. She was under the command of Captain Jefferson D. Howell, who, incidentally, was a brother-in-law of Jefferson Davis, the ex-President of the short lived Confederate States of America. On board were well over two hundred and fifty people and nearly $100,000 in treasure; much of it in gold and silver specie and gold bullion.

CAPTAIN JEFFERSON D. HOWELL

In command of the PACIFIC when she went down was Captain Jefferson D. Howell. He, along with nearly all the others on board, was lost with the ship.

CAPTAIN CHARLES A. SAWYER

Captain Charles Sawyer, the captain of the ORPHEUS.

On board the doomed ship hard-bitten miners from the gold creeks of the north country were thrust into close contact with some of the finest of Victoria's social elite; men of prominence like F. Garesche, of Garesche, Green and Company, bankers of Victoria and others like S. P. Moody of Moodyville and Captain Parsons of Fraser River fame. Below decks the inevitable games of chance began; careless miners with their pokes of gold and incautious businessmen pitted against the sharpened skills of professional gamblers; the ever present tinhorns who plied

The first leg of the trip passed uneventfully, then, just as the sidewheeler emerged from the Strait of Juan de Fuca and was rounding Cape Flattery, a promontory off the coast of Washington, she collided with the ORPHEUS, a sailing ship northward bound for coal. The initial impact, according to the survivors, was slight and almost impercepible, but the rotten hulk of the PACIFIC was breached. Almost immediately the vessel began shipping water at an alarming rate and within minutes she was listing badly. Strangely, the ORPHEUS, under the command of Captain Charles A. Sawyer, continued on her way without making any attempt to aid the stricken PACIFIC or those on board the imperiled ship. Captain Sawyer was later bitterly denounced for his callous disregard over the fate of the PACIFIC'S passengers. The sea was calm at the time of the mishap and he and his crew could easily have rescued the majority of the people on board the PACIFIC. It was, however, almost unprecedented when the ORPHEUS herself was wrecked and sank just off Cape Beale a few hours later. Nature had dealt out her own retribution but it was small consolation.

The impact brought most of the passengers and crew of the PACIFIC to the top deck, but when the ORPHEUS continued to sail northward the now panic stricken people realized that the other ship was not going to make any attempts to rescue them. The PACIFIC was now sinking very

Neil Henley, one of only two survivors from the ill-fated PACIFIC.

rapidly, so quickly, in fact, that there was not even time to launch her lifeboats. Suddenly, and almost without warning, the sidewheeler heeled over and slipped beneath the waves.

It had happened so swiftly that of all those on board the PACIFIC only two were eventually saved; a passenger named Henry Jelley, and Neil Henley, a quartermaster. One of the most tragic incidents in connection with the disaster was that which befell Miss Fanny Palmer, the daughter of the Digby Palmers of Victoria. In gay spirits earlier in the day, she had left for San Francisco but a few hours later her lifeless body was washed up on the beach only yards from her home. Nearly every family in Victoria had lost either a relative or an acquaintance in the calamity.

Nearly one hundred years have passed since the tragedy occurred but the rotting remains of the sidewheeler still lie in only tuelve or thirteen fathoms of water off Cape Flattery. A mere handful of dedicated treasure hunters are familiar with her history and they are aware that her cargo, valued at only $100,000 at the time, is now worth somewhere around $700,000. The gold and silver specie, undoubtedly consisting of both Canadian and American coins - probably including many double Eagle, Eagle and half-Eagle gold pieces and Queen Victoria silver coins would average at least ten times their original face value. The gold bullion, much of it coarse or in nugget form, would fetch six or seven times as much as it did in 1875.

The storm tossed ocean still conceals her final resting place but someday, some enterprising diver will come across the debris of an old sidewheeler off Cape Flattery at a depth of twelve fathoms - and if he finds a gold coin it will probably be American with a date on it prior to 1875, then he'll know that he has found the long sought wreck of the PACIFIC, the bullion laden ship which has eluded treasure hunters for almost a century.

HANCOCK HOUSE – INDIAN TITLES

Argillite
Drew & Wilson
ISBN 0-88839-037-8 HC 40.00

American Indian Pottery
Sharon Wirt
ISBN 0-88839-134-X SC 3.95

Art & Style of Western Indian Basketry
Joan Megan Jones, PhD
ISBN 0-88839-122-6 SC 7.95

Art of the Totem
Marius Barbeau
ISBN 0-88839-168-4 SC 6.95

Artifacts of the NW Coast Indians
Hilary Stewart
ISBN 0-88839-098-X SC 12.95

Blackfeet
John C. Ewers
ISBN 0-88839-170-6 SC 7.95

Coast Salish
Reg Ashwell
ISBN 0-88839-009-2 SC 6.95

Haida. Their Art & Culture
Leslie Drew
ISBN 0-88839-132-3 SC 7.95

Hunters of the Buffalo
R. Stephen Irwin, MD
Illustrations by J. B. Clemens
ISBN 0-88839-176-5 SC 3.95

Hunters of the Eastern Forest
R. Stephen Irwin, MD
Illustrations by J. B. Clemens
ISBN 0-88839-178-1 SC 3.95

Hunters of the Ice
R. Stephen Irwin, MD
Illustrations by J. B. Clemens
ISBN 0-88839-179-X SC 5.95

Hunters of the Northern Forest
R. Stephen Irwin, MD
Illustrations by J. B. Clemens
ISBN 0-88839-175-7 SC 3.95

Hunters of the Sea
R. Stephen Irwin, MD
Illustrations by J. B. Clemens
ISBN 0-88839-177-3 SC 3.95

Indian Art & Culture
Della Kew & P. E. Goddard
ISBN 0-919654-13-4 SC 9.95

Indian Artifacts of the NE
Roger W. Moeller
ISBN 0-88839-127-7 SC 3.95

Indian Healing
Wolfgang G. Jilek, MD
ISBN 0-88839-120-X SC 11.95

Indians of the NW Coast
D. Allen
ISBN 0-919654-82-7 SC 3.50

Indian Quillworking
Christy Ann Hensler
ISBN 0-88839-214-1 SC 7.95

Indian Rock Carvings of the Pacific NW Coast
Beth Hill
ISBN 0-919654-34-7 SC 4.95

Indian Tribes of the NW
Reg Ashwell
ISBN 0-919654-53-3 SC 6.95

Indian Weaving Knitting Basketry of the NW
Elizabeth Hawkins
ISBN 0-88839-006-8 SC 3.50

Iroquois. Their Art & Crafts
Carrie A. Lyford
ISBN 0-88839-135-8 SC 9.95

Kwakiutl Legends
Reg Ashwell
ISBN 0-88839-230-3 HC 9.95

My Heart Soars
Chief Dan George
Helmut Hirnschall
ISBN 0-88839-231-1 SC 7.95

My Spirit Soars
Chief Dan George
Helmut Hirnschall
ISBN 0-88839-233-8 SC 7.95

N. A. Indians
Mike Roberts
ISBN 0-88839-000-9 SC 3.00

The Providers
R. Stephen Irwin, MD
ISBN 0-88839-181-1 SC 12.95

Spirit Quest
Carol Batdorf
ISBN 0-88839-210-9 HC 17.95

Those Born at Koona
John & Carolyn Smyly
ISBN 0-88839-101-3 SC 9.95

Tlingit. Their Art, Culture & Legends
Dan & Nan Kaiper
ISBN 0-88839-010-6 SC 4.95

Totem Poles of the NW
D. Allen
ISBN 0-919654-83-5 SC 3.50

We-Gyet Wanders On
Legends of the Northwest
ISBN 0-919654-99-1 HC 14.95

ESKIMO TITLES

Eskimo Life Yesterday
Hancock House
ISBN 0-919654-73-8 SC 3.95

Life with the Eskimo (Photo Essay)
Hancock House
ISBN 0-919654-72-X SC 2.95

The Incredible Eskimo
Raymond de Coccola
Paul King
ISBN 0-88839-189-7 SC 16.95

HANCOCK HOUSE – NATURAL HISTORY TITLES

Alpine Wildflowers
J. E. (Ted) Underhill
ISBN 0-88839-975-8 SC 4.95

Backroads Explorer: Similkameen & S. Okanagan
Murphy Shewchuk
ISBN 0-88839-205-2 SC 9.95

Birds of Alberta
David Hancock
ISBN 0-88839-222-2 SC 6.95

Birds of B.C.
David Hancock
ISBN 0-88839-218-4 SC 6.95

Introducing the Birds of North America
David Hancock
ISBN 0-88839-220-6 SC 6.95

Birdwatching in Canada
Sterling & Woodford
ISBN 0-919654-28-2 SC 3.95

Coastal Lowland Wildflowers
J. E. (Ted) Underhill
ISBN 0-88839-973-1 SC 4.95

Edible Seashore (cookbook & guide)
Rick M. Harbo
ISBN 0-88839-199-4 SC 5.95

Exploring the Outdoors/SW B.C.
Tony Eberts & Al Grass
ISBN 0-88839-989-8 SC 9.95

Falconry Manual
Frank Beebe
ISBN 0-88839-978-2 SC 12.95

Fur, Gold and Opals (Guide to Thompson R.)
Murphy Shewchuk
ISBN 0-919654-36-3 SC 3.95

Guide to Indian Herbs
Raymond Stark
ISBN 0-88839-077-7 SC 5.95

Guide to the Western Seashore
Rick M. Harbo
ISBN 0-88839-201-X SC 5.95

How to Catch Really Big Fish
Tara Robinson
ISBN 0-88839-967-7 SC 3.95

Indian Herbs
Dr. Raymond Stark
ISBN 0-88839-077-7 SC 5.95

Mooching: Salmon Fisherman's Bible
David Nuttall
ISBN 0-88839-097-1 SC 6.95

NW Roadside Wildflowers
J. E. (Ted) Underhill
ISBN 0-88839-108-0 SC 5.95

NW Wild Berries
J. E. (Ted) Underhill
ISBN 0-88839-027-0 SC 7.95

NW Wild Harvest
Terry Domico
ISBN 0-88839-022-X SC 7.95

Orchids of N. America
Dr. William Petrie
ISBN 0-88839-089-0 SC 9.95

Pacific Wilderness
David Hancock, Lyn Hancock & David Sterling
ISBN 0-919654-08-8 SC 4.95

Rafting in British Columbia
Doug VanDine & Bernard Fandrich
ISBN 0-88839-985-5 SC 6.95

Rocks & Minerals of the NW
Stan Leaming & Chris Leaming
ISBN 0-88839-053-X SC 3.00

Sagebrush Wildflowers
J. E. (Ted) Underhill
ISBN 0-88839-171-4 SC 4.95

Tidepool & Reef
Rick M. Harbo
ISBN 0-88839-039-4 SC 5.95

Upland Field & Forest Wildflowers
J. E. (Ted) Underhill
ISBN 0-88839-174-9 SC 4.95

Western Mushrooms
J. E. (Ted) Underhill
ISBN 0-88839-031-9 SC 3.95

Western Wildlife
David Hancock
ISBN 0-919654-81-9 SC 3.95

Wildlife of the Rockies
David Hancock
Photos: Tom W. Hall
ISBN 0-919654-33-9 SC 3.50